International Scientific Relations

International Scientific Relations

Science, Technology and Innovation in the International System of the 21st Century

Francisco Del Canto Viterale

ANTHEM PRESS

Anthem Press
An imprint of Wimbledon Publishing Company
www.anthempress.com

This edition first published in UK and USA 2021
by ANTHEM PRESS
75–76 Blackfriars Road, London SE1 8HA, UK
or PO Box 9779, London SW19 7ZG, UK
and
244 Madison Ave #116, New York, NY 10016, USA

Copyright © Francisco Del Canto Viterale 2021

The author asserts the moral right to be identified as the author of this work.

All rights reserved. Without limiting the rights under copyright reserved above, no part of this publication may be reproduced, stored or introduced into a retrieval system, or transmitted, in any form or by any means (electronic, mechanical, photocopying, recording or otherwise), without the prior written permission of both the copyright owner and the above publisher of this book.

British Library Cataloguing-in-Publication Data
A catalogue record for this book is available from the British Library.

Library of Congress Control Number: 2021942253

ISBN-13: 978-1-78527-707-8 (Hbk)
ISBN-10: 1-78527-707-3 (Hbk)

Cover image: By sdecoret/Shutterstock.com

This title is also available as an e-book.

To the memory of my beloved father, Francisco Del Canto Gonzalez, who was always an excellent model of love, respect, and honesty to me.

CONTENTS

Foreword by Fernando López-Alves   xiii
Preface   xvii
Acknowledgments   xxiii

PART 1.   INTRODUCTORY FRAMEWORK

1. Science, Technology, and Innovation and International Relations   3
   1.1 Science, Technology, and Innovation   4
      The Dawn of Science   4
      Science, Technology, and Innovation as an Empirical Phenomenon   5
      Science, Technology, and Innovation as a Subject of Study   7
   1.2 Science, Technology, and Innovation and International Relations   8
      Practical Links   8
      Disciplinary Approach   10
   1.3 Rise of Science, Technology, and Innovation   13
      New Role of Scientific Knowledge   13
      Knowledge Economy   14
      Knowledge Society   16
   1.4 International Scientific Relations   17
      New Empirical Phenomenon   17
      New Subdisciplinary Field   18
      Challenges for International Scientific Relations   19

PART 2.   ANALYTICAL FRAMEWORK

2. International Context   23
   2.1 *Actors of the International System*   23
      Changes in the Role of the Nation-State   24
      New Actors   24
      Multipolar World   25
   2.2 *Interactions and Relations within the International System*   26
      Interdependence   26
      New Global Agenda   27
      Types of Linkages   28

	2.3 *Main International Processes*	30
	End of the Cold War	30
	Challenges to the Nation-States	31
	Scientific and Technological Revolution	32
	Changes in the Capitalist Economic System	34
	Globalization	35
	2.4 New Configuration of the International System	38
	Intersystemic Transition	38
	Towards a New Global Configuration	39
3.	Actors	41
	3.1 Old and New Actors	41
	3.2 Universities	42
	Historical Particularities	43
	New Roles and Challenges	45
	New Actors	45
	Autonomy	46
	New Roles	46
	Reconfiguration and Adaptation	47
	3.3 Nation-States	48
	The Emergence of Big Science	49
	Post–Cold War Era	50
	Sub-State Entities	52
	3.4 Intergovernmental Organizations	53
	Specialized Organizations	54
	Regional Processes	55
	3.5 Nongovernmental Organizations	57
	Rise of the NGOs	57
	Promotion and Uses of Science, Technology, and Innovation	58
	Impulse and Promotion of Science, Technology, and Innovation by NGOs	58
	Use of STI as a Tool	59
	Main NGOs in Science, Technology, and Innovation	59
	3.6 Transnational Companies	60
	Economic Interest	61
	Action Strategy	61
	3.7 Think Tanks	63
	Predecessors	63
	Distribution and Geographical Expansion	64
	Margins of Autonomy	65
	3.8 Epistemic Communities	65
	New Global Context	66
	Revaluation of Expert Knowledge	67
	3.9 Scientific Diasporas	68
	From Brain Drain to Brain Gain	69
	Main Initiatives	69

CONTENTS ix

4. Relations — 71
   - 4.1 Interaction and Relation Mechanisms — 71
   - 4.2 Conflict Interactions — 73
     - Fighting for Global Talent — 73
     - *Historic Evolution* — 75
     - *Lack of Global Talent* — 75
     - *Public Policies* — 76
     - *Impact* — 78
     - Intellectual Property Rights and Patents — 79
     - *Origin* — 79
     - *Vertiginous Evolution* — 80
     - *Clash of Interests* — 81
   - 4.3 Cooperation Dynamics — 82
     - International Cooperation — 83
     - STI Diplomacy — 85
     - Links between State, Company, and University — 87
     - Interuniversity Cooperation — 89
   - 4.4 Competitive Interactions — 90
     - Competitive Race between States — 90
     - Competition between Companies — 91
     - University Rankings — 93
     - Fight for New Emerging Technologies — 95
   - 4.5 Asymmetric Relations — 97
     - Historic Inequalities — 97
     - Power and Knowledge — 98
     - Cognitive Divide and Dependency — 99

5. Processes — 101
   - 5.1 Production — 102
     - Methods of Production — 102
     - *Traditional Methods* — 103
     - *New Methods of Production* — 104
     - *Alternative Methods* — 107
     - Increase in the Factors of Production — 108
     - *Research and Development (R&D)* — 109
     - *Researchers* — 110
     - *Higher Education* — 110
     - *Infrastructures* — 112
     - *Publications and Patents* — 113
   - 5.2 Intermediation — 114
     - Transmission through Formal Education — 115
     - *Higher Education* — 116
     - *Scientific Research* — 116
     - *Lifelong Learning* — 117

Cooperation Networks	118
E-learning	119
Knowledge and Technology Transfer	121
Increasing Relevance of Transference	121
Characteristics of the Process	122
Diffusion through Information and Communication Technologies	124
Dissemination by Mass Media	125
Development of Digital Media	125
Mobility and Circulation of Highly Skilled Personnel	128
Movement Dynamics	128
Global Impact	129
5.3 Distribution	130
Geopolitical and Geoeconomic Distribution	130
R&D Investment	131
Researchers	133
Geography of Scientific Publications	134
Patents	135
Academic Institutions and Postgraduate Students	136
World-Class Universities	137
International Students	138
5.4 Application	138
Type of Research	139
Scientific Fields	140
Publications and Patents	141
Socioeconomic Impact	142
Private Sector	143
5.5 Governance	144
Public Policies	145
Local Governments	148
Regional Initiatives	150
New International Sphere	151

PART 3.   EXPLICATIVE FRAMEWORK

6. Emerging Realities	157
6.1 New Emerging Phenomena	158
6.2 Knowledge Gap	158
Types of Gaps	158
Digital Divide	159
Cognitive Divide	160
Scientific Divide	160
Impact of the Divides	161
New Centers and Peripheries	161
Intrasocietal Differences	161
Reinforcing Old Social Divides	162

	6.3 Multilevel and Global Governance	162
	Multilevel Governance	163
	Global Governance	164
	Future Challenges	165
	6.4 Commercialization of Science, Technology, and Innovation	166
	Commercial Value of Science, Technology, and Innovation	166
	Economic Models of Knowledge Production	167
	Decline of Societal Relevance	168
	6.5 Gender Divide	169
	Women in Science, Technology, and Innovation	169
	Ambivalent Trends	170
	Public Policies	171
	6.6 Geopolitical Changes	172
	Regional Changes	172
	China's Rise	173
	Geopolitical and Geoeconomic Shift	174
	6.7 Science and Emerging Technologies	175
	New Scientific and Technological Areas	176
	Future Implications	177
	6.8 Virtual World	180
	Development and Cooperation	180
	Virtual Conflicts	181
	Cyberspace Governance	182
7.	Roles of Scientific Knowledge	185
	7.1 Resource for Economic Gain	185
	7.2 Instrument of Power	187
	7.3 Mechanism of Social Innovation	189
	7.4 Democratizing Element	192
	7.5 Strategic-Military Factor	194
8.	Configuration of International Scientific Relations	199
	8.1 Systemic Parameters	199
	Actors	199
	Interactions and Relations	201
	Processes	201
	Production	202
	Intermediation	202
	Distribution	203
	Application	203
	Governance	204
	Emerging Realities	204
	8.2 Global Structure	205
	Polarization	205
	Distribution	207
	Hierarchization	208

	Segmentation	209
	Sectorization	210
9.	International Scientific Relations and the International System	211
	9.1 Interactions with Other Subsystems	212
	Economic System	212
	Political System	213
	Strategic-Military System	214
	Social System	215
	9.2 Macrotrends	216
	Continuity Macrotrends	217
	Change Macrotrends	217
	9.3 Implication for the International System	218
Conclusions		223
	International Scientific Relations and International System	230
	Epilogue	231
References		233
Index		247

FOREWORD

For the first time in the history of humanity, we bear witness to a time where the international system is increasingly dependent on a large number of actors. The past 300 years of history clearly show that we are seeing something unusual: *the continuous multiplication of actors capable of affecting this international system*. The Nation-State has lost its central role as the sole sovereign actor of the international system and must now interact and negotiate with new actors such as NGOs, mafias, multinational corporations, and hackers, among many others.

To this multitude of actors and interactions, we also add a new spectacular revolution in communications (a term too old to capture what is truly happening nowadays, with all its nuances and ramifications), which is introducing a new way of building power and a different way of establishing alliances and creating conflicts on an international scale.

These two factors have generated a new international scenario in which the actors, both large and small, have equal access to state-of-the-art technology (military, biological, energetic, chemical, industrial, and cybernetic), which gives these domestic- and international-level entities a substantial increase in power and influence. We are already three decades into what has been called the *computerization* of armies, the *automation* of air armaments, the implementation of *high tech* to the navy, and the rise of long-distance wars. These advances, along with cyberterrorism, make States much more dependent on access to new technologies now than in any previous point in history.

Unlike what happened in the past, nowadays science, technology, and innovation are changing rapidly and have a decided influence on global power. This combination has created a new dimension of something that is already very well known: *scientific and technological espionage*. The global system is no longer an arena totally controlled by States and, as such, espionage is no longer an activity funded and dominated completely by governments.

As has always happened, Nation-States live with the constant pressure of trying to improve the technology linked to its military apparatus. However, scientific and technological development is not only reduced to traditional armament but also biological and chemical weapons, nuclear technology, communicational and visual know-how for high-precision large-distance warfare and cybernetic combat. Nowadays, the possession of nuclear weapons isn't the be-all-end-all; on the contrary, illicit access to information (hacking) has made it possible for small and medium actors to challenge the stronger ones through mechanisms that can create confusion, generate political instability, and undermine politicians' legitimacy.

The public investment needed for all this to work has tripled in the last two decades, at least among world powers. In terms of expenses, the number of people and resources governments must employ to not fall behind in this competitive race has quadrupled (China and the United States are spearheading these efforts). At the same time, States are constantly on watch, trying to block the access of other actors to the scientific and technological advances that might give them a competitive edge. Another interesting indicator is that, in the last decade of the 20th century and in the 21st century, we saw an extraordinary increase in the number of registered patents, most of all presented by the traditional actors (again, the United States, China, and Russia), but there is also an increase in other peripheral zones of the planet.

The contemporary international system is much more complex and unstable than the world order that ended in 1989 with the end of the Cold War and the decades following it. Today's superpowers are forced to create new negotiation mechanisms (even with small- and medium-sized powers) that did not exist twenty years ago. These peripheral countries are peripheral no longer and many of them can rapidly transform into threatening actors. For example, the States that lead the "nuclear club" have been forced to accept new members that now claim a seat at the negotiation table and want to speak and vote at the United Nations Security Council.

To summarize, those at the peak of the global system (the United States, China, and Russia at the top; Japan, Germany, England, and India one step below; and, lower still, France, South Korea, Canada, and Indonesia) are waging among themselves a war of influences characterized, on the one hand, by scientific and technological espionage and, on the other hand, by the development and funding of the scientific enterprise. The way in which scientific research is carried out, who is funding it, what types of projects are considered a priority, and how these research institutions associate themselves internationally and politically, add a further layer of complexity to the relationship between science, technology, and innovation and international relations.

Lastly, we can consider that we are dealing with a new international system in which the characteristics of global economy and individual States depend, more than ever before, on the access to science, technology, and innovation.

Even though the first scientific and technological revolutions (sedentary agriculture, steel, and the so-called industrial revolution of the 19th century) only affected a small part of the world's population, in the last two decades we are witnesses to an important change produced by a new scientific and technological revolution (which includes artificial intelligence and the creation of new algorithms that have allowed the development of new information platforms that would have been unthinkable years back, such as Netflix and Facebook) that is radically life-changing for the world's entire population.

Because of that, the current international system can't be compared to the previous ones, nor does it have a direct continuity with them; on the contrary, as argued by Richard Haass, our current system is a *new world disorder* rather than a new world order, also understood as *System 0*, where not even the ideas of liberalism or neoliberalism are enough to explain what kind of system is emerging. In truth, what has consolidated alongside the scientific and technological race is an unstable, individualist and competitive political system, where protectionism and nationalism have conquered the core of the system, taking root in Europe, the United States, Asia, and a large part of Latin America.

Because of this global context, it would be very naïve to think that the phenomenal change in science and technology and the way that scientific research is now done do not have a profound impact on international relations. This book by Prof. Francisco Del Canto Viterale seeks to give an answer to some of the questions about the new reality of knowledge, science, technology, and innovation in this international system, thanks to a systematic analysis that tries to elucidate the why, the how, and the where of all these processes.

As argued by the author of this book, science, technology, and innovation is, now more than ever before, linked to the national destinies and the interactions between international actors, and the slightest alteration in the scientific and technological *balance of power* on a global level can create a completely new international system. Because of that, there is no doubt that the subjects broached in the following pages are not only relevant but decisive for the future world order of the 21st century.

Fernando López-Alves, PhD
Professor of Sociology, Global,
and International Studies
University of California,
Santa Barbara, CA, 93106.

PREFACE

Being witnesses to an extraordinary historical moment due to the quantity, depth, and complexity of the changes that are taking place, the search for valid explanations that give an answer to new topics, phenomena, and actors that can be observed in the current international system must be understood as a social and scientific need. The starting point for this book has been the author's deep belief in the urgency of advancing the research on the intense changes that have taken place in the last few years in the field of international studies. The strong restructuring of the world order and the magnitude and speed of these changes make analyzing and understanding these global transformations, as well as rethinking the theoretical and methodological framework we use to give an answer to the phenomena we study, an unavoidable task.

The profound modifications that have taken place at the core of the international system in the past 30 years have given rise to a new and complex world configuration characterized by the growing phenomenon of globalization, the intensification of the scientific and technological revolution, the appearance of new international actors, and the emergence of a renewed global agenda. In the tumultuous global context of the beginning of the 21st century, *science, technology, and innovation* (STI) acquires a special and strategic relevance as a global agenda topic. Even though scientific knowledge has been an important factor in many past societies, what has changed is that it has become indispensable in this new phase of international relations, as the main source for generating economic benefits, political power, and social development.

The links between STI and international relations are not a new phenomenon in the international system; on the contrary, scientific knowledge has always been considered an important factor in many historical periods, from Ancient Egypt, through the European colonial empires, to its peak in the 20th century. However, in the current international system, the decisive factors are the magnitude, depth, and intensity those links have acquired. The world is changing rapidly and intensely, and scientific knowledge is one of the main causes, having become a critical variable in the international system where its prominence and influence are only growing.

The analysis of the post–Cold War international system allows us to verify the birth of a new world order, still uncertain in many of its parameters, but where we can still identify the more dominant role of scientific knowledge. This has resulted in the expansion of a new empirical area at the intersection between scientific and technological topics and international affairs called *international scientific relations (ISR)*. Precisely, although it might be thought that STI has had concrete influence all through the history of international relations, its revaluing in the second half of the 20th century and the start of the 21st century have transformed ISR into a subject of study that requires more attention from the academic field.

Considering this context, the main objective of this research is to produce a detailed, deep and exhaustive analysis of the new reality of STI in the international system, as well as to discuss the main changes and continuities that are currently taking place in the subfield of ISR. With a systemic and methodological approach, we intend to find accurate and novel answers, explanations, and conclusions on three main points:

(i) To discover the *role played by STI* in the current international system.
(ii) To analyze the main characteristics of the current *global configuration of ISR*.
(iii) To explain the *impact that ISR* has on the international order.

The research analyzes all these changes through an innovative and rigorous methodological design with the following characteristics:

- This research is attached to the field of social sciences, particularly to international studies and specifically to the subdiscipline of *international scientific relations*.
- Complementarily, and due to the complexity and the multidisciplinary nature of STI as a subject matter, the research uses an interdisciplinary approach, using information, points of view, analyses, and methods from many scientific disciplines.
- The systemic perspective is the main methodological strategy. It is understood that ISR can be studied and analyzed as a complex system, with both individual and global properties, that acts as a whole, due to the fact that its components interact and are tightly linked together. Through this vision, the goal is to integrate the analysis of micro and macro levels

while also carrying out an analytical study of its parts and a holistic study of the system in its entirety.[1]
- An original systemic model, created by the author, is applied to the international stage to analyze and explain ISR. The systemic model is a methodological analysis tool that, in a simplified and, as such, comprehensible fashion, tries to represent some aspects of reality and turn the subject matter into a more easily understandable phenomenon.[2]
- For the data collection and data analysis, it was used a large variety of statistical sources (including all major international organizations) and an extensive literature review (from the fields of international studies, other areas of social sciences, and other technical disciplines).

Using this methodological design, the goal is to offer new answers and explanations in the field of social sciences, specifically in the discipline of international studies, about the role of STI in the international system and about the new dynamics of ISR.

The reasons that justify the creation of a work such as this are multiple, diverse, and interconnected. The first reason is related to the deep *structural changes* that the international context has gone through in the past 30 years, which need new academic approaches and scientific explanations that give an account of the actors, phenomena, and processes that are emerging on a global scale. Within this renewed international framework, one of the most impactful issues in the new global agenda is the scientific and technological revolution that has given an extraordinary boost to STI and is having perceivable consequences on the global system as a whole. The 21st century's new international context, changing, innovative, and unknown, is precisely the one stimulating the search for new answers.

The second is a *thematic* reason linked to the revaluation of STI as a subject matter. Even though scientific knowledge has been a relevant research topic throughout history, it is also true that the scientific and technological developments of the past few decades make it necessary to adopt new theoretical and methodological approaches that are aware of the new reality of the phenomenon. This approach requires advances in research lines that give proof of the changes that have taken place in STI, thus proving that theory must go together with reality. At the same

---

[1] System is defined as "a totality made up of several elements that all interact with each other, which is distinguished from its environment or medium, from which it receives a wide range of stimuli, across a border and which constitutes a distinguishable unit" (Bunge, 1980a, 1996, 2009).

[2] The system model applied in this investigation to analyze ISR was designed by the author and it was published as a separate scientific paper (Del Canto Viterale, 2019).

time, much of the research on this topic (including that conducted from the lens of international studies) has had as its principal characteristic an atomized and fragmented approach of the phenomenon, focusing solely on particular aspects of STI, which exposes the need of an approach oriented to the search for more holistic explanations of the phenomenon.

The third is *disciplinary* reasons that justify this research. Currently, international studies are going through a period of revision and reconversion of their academic foundations, with the goal of better interpreting the international reality through the study of new subjects, approaches, and methods. If we consider the specific subfield of ISR as an expanding subdiscipline, it is expected that the treatment given here to scientific knowledge as a subject of study can be a helpful contribution to analyze this new issue.

There are also *methodological* reasons that justify this research. The existence of new and complex empirical phenomena in the international reality forces researchers to develop new mechanisms, tools, and instruments, which can be used to validate scientific hypotheses and experiments in order to approach the subjects of study. The systemic perspective, in force since the beginning of the 20th century, is having a renewed boom because of its interest in finding integral answers and holistic understanding about complex phenomena. Because of that, the usage of the systemic approach can offer an interesting methodological approach to study international events such as STI in order to understand them to their full extent.

The present research carries out a deep and detailed systemic analysis that studies the parameters, global configuration, and role of STI in ISR, and the implications this has on the international system as a whole. To fulfill these expectations, the present book has been divided into three parts:

1) In Part 1. Introductory Framework, we define and describe the main concepts studied in the book, and we analyze the factual and theoretical evolution of the links between STI and international relations (Chapter 1. "Science, Technology, and Innovation and International Relations").

2) In Part 2. Analytical Framework, we present four chapters where each of the parameters that make up ISR is systematically studied. To do so, we describe the changes undergone by the international system, understood as the conditioning environment for STI (Chapter 2. "International Context"); we identify and analyze the distinct international actors related to scientific knowledge (Chapter 3. "Actors"); we study the main interactions and relations established within ISR (Chapter 4. "Relations"); and, lastly, we examine scientific knowledge itself, seeking to understand all the processes and operative mechanisms it generates in the international

system from its creation and intermediation to its governance and final application (Chapter 5. "Processes").

3) In Part 3. Explanatory Framework, we introduce the last four chapters of the book, where we analyze the new characteristics that STI has acquired within the frame of ISR in the 21st century, and we give answers to the three objectives established at the beginning of the research project. To do so, we identify the new realities and phenomena emerging in ISR (Chapter 6. "Emerging Realities"); we evaluate the changes in the functions and uses of STI in the international system and the new roles it plays in the current world order (Chapter 7. "Scientific Knowledge's Roles"); we study the main characteristics of the current global configuration of ISR (Chapter 8. "Configuration of International Scientific Relations"); and, finally, we describe the impact and main consequences of ISR on the current international system (Chapter 9. "International Scientific Relations and the International System").

Figure 1 is shown to the reader with a double goal: one is to introduce the systemic model created by the author and used in this research for the analysis of ISR, and the other is to present a roadmap to help the reader understand the book. We hope that this visual tool is able to offer a quick and simple answer about the interaction between the systemic perspective used as a methodological tool

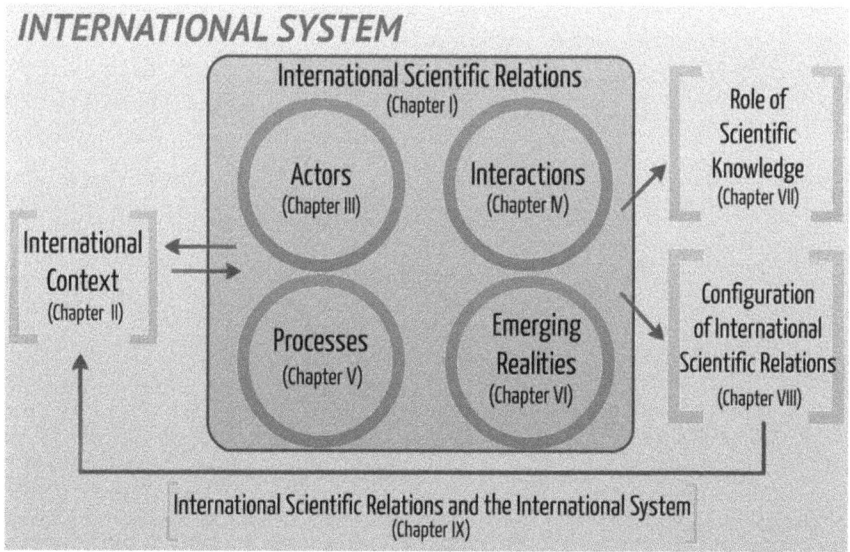

Figure 1. Road Map

and the analysis of the subjects approached in the research, all while helping follow the research by linking the chapters with its subjects.

In short, the reader is presented with an ambitious intellectual work with regards to the objectives it pursues, but which is also aware of the difficulties and limitations it faces when analyzing a problem as complex as the link between STI and international relations. To face this challenge, the author has chosen to carry out a research project focused on a subject of study (science, technology, and innovation), belonging to a subdisciplinary field (international scientific relations), using a methodological approach (systemic perspective), and applying research techniques (systemic models) that can offer great analytic potential to the field of international studies and social science.

The current empirical and theoretical complexity faced by international studies is not unknown, nor is the need for new approaches that link STI to international relations; on the contrary, this state of uncertainty and indefiniteness is what encourages the development of the following research work seeking to offer answers to a disciplinary field and a social knowledge that requires it.

ACKNOWLEDGMENTS

I want to extend my gratitude to all those who have collaborated in the making of this book. To the universities where I have studied and worked: University of Salvador (Argentina), FASTA University (Argentina), Universidad of Deusto (Spain), Universidad of Salamanca (Spain), Johns Hopkins University (United States), Massachusetts Institute of Technology (United States), University of North Dakota (United States), and Fresno Pacific University (United States). To the professors and researchers who have so generously accompanied me on this long road and have become true mentors through their professional and human advice. To all those friends who turned my stays in Mar de Plata, Buenos Aires, Bilbao, Baltimore, Boston, Fresno, and Salamanca into beautiful experiences and life lessons. To my parents, my family, my friends, my city, and my country. To Manuela and Isabella. To all of them, many thanks for helping me fulfill this beautiful dream.

Part 1

INTRODUCTORY FRAMEWORK

Chapter 1

SCIENCE, TECHNOLOGY, AND INNOVATION AND INTERNATIONAL RELATIONS

Although the links between science, technology, and innovation (STI) and international relations have been very close throughout history, the magnitude, depth, and relevancy acquired by this relation in the current international stage is unprecedented. The deep changes that have taken place in the international system in the past 30 years have boosted STI as a key instrument for achieving economic, political, and social development. This has resulted in the acceleration and intensification of the interaction dynamics between STI and international relations with strong repercussions for the configuration of the world order.

The scientific and technological transformations and their economic, social, political, and cultural implications that have occurred within the international system from the last quarter of the 20th century to the present day give the impression of living in a key moment in the history of humanity. The speed and depth of the intensive application of new technology to numerous areas of daily life make it difficult to assess the impact that all these changes are generating on the new international system. Nonetheless, it is safe to say the post–Cold War international system is experiencing a reconfiguration of its main systemic parameters in which STI has become a critical factor.

Despite the historic relevance of STI to the international system, in the theoretical and academic field of international studies, the treatment of the phenomenon of scientific knowledge has been rather sporadic, reduced, and unarticulated. This growing and strategic link between STI and international relations has driven the social and academic need to examine and deepen the analysis of how scientific knowledge and international affairs connect with, impact, and provide feedback to each other.

With this objective in mind, this introductory chapter will address some basic conceptual definitions, a brief evolution of STI throughout the history of international relations, a description of STI in a new context of *knowledge economy* and *knowledge society*, and introduce the novel interaction and overlap between STI and international relations,

1.1 Science, Technology, and Innovation

The dawn of science

The origin of scientific knowledge and modern science can be traced back to Francis Bacon's *Novum Organum* (1620) and Descartes's *Discourse on the Method* (1637), the starting point from which we can consider knowledge to be scientific.[1] The key point was to understand that the world is objectively understandable and can be captured with concepts and representations built by reason. From this radical separation between the subject and object, rooted in Christian thought and secularized by Descartes, the construction of knowledge is objectivized and can, as such, be generalized and transformed into universal. In this sense, modern science can be seen as a vast enterprise that uses a large amount of human effort to acquire concrete knowledge about reality and that can be understood as an activity and a product of our culture, which has allowed humanity access to more systemic and rigorous knowledge: scientific knowledge (Bunge, 1972, 1979, 1996, 2009).

Francis Bacon himself, who laid down the theoretical foundations of what we consider modern science in his *Novum Organum* (1620), distinguishes three main aspects in it: (i) the *material and human resources* needed to generate knowledge, (ii) the *method* of acquiring and generating new knowledge, and, finally, (iii) *scientific knowledge* itself (Lamo de Espinosa, González García, and Torres Albero, 1994).

(i) The first perspective sees science as a *social institution*, with a growing autonomy from other institutional sectors such as economics, politics, or education, formed by organized workers and with certain material and economic resources.

(ii) The second aspect, the *scientific method*, is established as the validity criterion for scientific knowledge. This is understood as science's characteristic procedure for dealing with a series of issues, and it is what allows us to analyze and systematize information through the process of scientific research (Bunge, 1980b, 1996, 2003). As Ander-Egg (2001, 2010) points out, we can understand that scientific research "is the process that, through the scientific method, allows to obtain new knowledge in the

---

[1] Vulgar, artistic, philosophic, or religious knowledge can be considered of equal relevance and interest as scientific knowledge but are not part of the subject of study of this investigation.

field of social reality (pure research), or to study a situation to diagnose needs and issues with the goal of applying said knowledge to practical goals." In essence, the underlying idea is considering the process of scientific research as the most adequate, methodical, and productive way of creating useful and reliable knowledge for society.

(iii) The third dimension pointed out by Bacon is the *scientific knowledge* itself, understood as the *final product* of the research process carried out with the help of the scientific method. As Ander-Egg (2001, 2010) explains, scientific knowledge is born because humans, who "naturally want to know" (Aristotle), are neither satisfied with perceiving what is external, nor with common sense, because there are phenomena that are not understood on a single perceptive level and compel us to go beyond ordinary knowledge; this jump leads to scientific knowledge. Between one type of knowledge and the other, there is a degree of division, because the difference is not in the nature of the object of study but in the procedural method of the acquisition of knowledge.

If the three dimensions of science (the social institution, the method, and scientific knowledge) are analyzed, scientific knowledge itself is the most relevant socially, due to the strong impact it generates in many aspects of human and social life.

Science, technology, and innovation as an empirical phenomenon

Science, technology, and innovation have been strongly interconnected among themselves, in particular, in the past few decades, and this increasing interaction has led to the generation of a unique empirical phenomenon characterized by being complex, multicausal, and multidimensional. At the same time, it is also possible to describe them as individual phenomena that present their own particularities and each has evolved independently throughout international relations history.

Science is understood as the human enterprise tasked with producing authoritative and systemic knowledge to understand the world that surrounds us. Through science, humanity has built an ordered body of knowledge, obtained through observation and experimentation in specific fields and organized and classified on the basis of explanatory principles. In recent years, the term *science* has also been associated with *pure or basic science*, which is the kind of research with no goal other than acquiring knowledge by itself, excluding practical interests, but which, at the same time, is the theoretical foundation of knowledge on which applied science or technology is based (Bunge, 1980b, 2003).

Then, *technology* is science applied to resolving specific issues. It is a body of scientifically ordered knowledge which allows us to design and create goods or services that help with adapting to the environment and the satisfaction of the basic needs and wishes of humanity. We can consider technology to be the means by which scientific knowledge is applied to the solution of problems using devices, procedures, tools, and skills (Van Wyk, 2004). As Mario Bunge (1972) points out, "Science as an activity (such as research) belongs to social life; insofar as it is applied to the betterment of our natural and artificial environments, to the invention and manufacture of material and cultural goods, science becomes technology." Technology has been responsible for humanity's great leap in the last centuries, which has allowed a rural society to transform into a complex modern society, with significant progress made in practically all fields.

Lastly, *innovation* is a process of economic value generation through which certain products or production processes, developed with new knowledge or the innovative combination of pre-existing knowledge, are effectively introduced in markets and social life.[2] It is essentially pieces of knowledge that are combined, applied, and distributed in a novel way in the processes of interaction and learning between different actors in the frame of cooperative relations and innovation networks on a territorial and global level. The Oslo Manual understands it as "the implementation of a new or significantly improved product (good or service), or process, a new marketing method, or a new organizational method in business practices, workplace organization or external relations" (OECD and Eurostat, 2006; OECD, 2018). What really differentiates *innovation* from *invention* or scientific *discoveries* is the valorization and commercial utility of knowledge generated through the production of a demand for new goods or products in the market.

At present, these conceptual variations do not translate into significant differences when analyzing their impact on international relations. Although, as described, science, technology, and innovation have their own particular characteristics, they must be considered as elements within a singular system that jointly form a single and unique empirical phenomenon. They are differentiated but interconnected phenomena within the same process, where different scientific activities interact to form a complex system that transforms discoveries and inventions in technologies, innovation, and scientific knowledge with specific applications in different areas of society. For this reason, it is possible to consider all these processes to have transformed into the new and

---

[2] Joseph Schumpeter (1961) introduced the concept of *innovation* in his book "Theory of Innovations," where he defines it as the factors of production combined in a novel way and that are the key to economic growth.

indivisible empirical phenomenon within the current international system we call *science, technology, and innovation (STI)*.

STI is to be understood as a set of scientific tasks that interact and are linked between themselves and with other socio-economic dimensions in the international system and that have great consequences and implications for the actors, relations, processes, parameters, and global configuration of the world order. It is, therefore, an empirical phenomenon that has transformed into a significant dimension of the 21st century's international relations.

Science, technology, and innovation as a subject of study

Most scientific disciplines have approached scientific knowledge as a subject of study. The majority of those disciplines have tried to address the phenomenon with very different approaches and focuses but always considering its relevance as an element of scientific analysis. For instance, philosophy, history, sociology, economics, and political science, among others, have developed a deep and substantive analysis of STI. Scientific knowledge has also driven the creation of a new disciplinary field that has taken it as its subject matter like the case of science, technology, and society.

In the beginning, scientific knowledge and especially the advances in science and technology were the subject of study of the more technical disciplines that were directly involved with the objective and mechanistic concepts that prevailed over them. However, the changes and the increasing impact of the scientific and technological developments in social contexts allowed disciplines in the field of social sciences to carry out a deeper and more critical study of the phenomenon of scientific knowledge.

As the role and impact of STI grew as an empirical phenomenon, so too did its theoretical treatment and, broadly speaking, it is possible to identify two opposing perspectives that have prevailed in the field of social sciences about the way of understanding the impact of science and technology on society:

> On the one hand, the *optimistic view* understands that the advances in science and technology tend to build a better society, which is reflected in the economic development and the betterment of the human condition and general well-being. This perspective assumes that STI allows us to solve major human challenges such as public health care, energy, food, water, poverty, or climate change. This perspective is shared by the majority of international organizations (such as the World Bank, United Nations, or OECD) and by a large number of experts and specialists (especially those in traditional paradigms).

On the other hand, there also exists a more skeptical perspective on the role of science and technology as a mechanism for solving society's problems. It is a *pessimistic view*, which understands that developments in science and technology have had (and will continue having) negative (and, in some cases, disastrous) consequences for humanity. In this vein, there are even some that see scientific and technological determinism as a strong modern ideology that is imposed over diverse cultural and economic contexts (critical theory and postmodernism approach). It is understood that science and technology cannot be considered neutral or impartial, and as proof of this are the many times where it has been used for colonial domination, racism, and exploitation.

These differences in the way of understanding the role and impact of science and technology in social sciences can also be observed within the field of international studies.

1.2 Science, Technology, and Innovation and International Relations

Practical links

Scientific knowledge has been an active part of humanity's history and its consequences have had implications not only on individuals and societies but also on the relations between peoples and nations. Science and technology have had a clear influence throughout world history since ancient Egypt, through Greece and the Roman Empire, to the development of Eastern civilizations, where scientific knowledge had a prominent role as a tool for economic and social progress and, at the same time, in the usage of new technological developments applied to the military sector. In fact, scientific and technological developments have been particularly relevant in explaining a large part of the history of international relations due to the new advances and inventions being decisive factors in the control of land and sea, thus helping the expansion of many empires all throughout the planet: the processes of colonization and conquest carried out successively by the Spanish, Dutch, French, and British empires between the 16th and 19th centuries can only be understood as a consequence of these scientific and technological developments and because of the application of these developments to the benefit of their empires.

As Paul Kennedy (1987) points out, the dynamics of technological change and military competitiveness was the great driving force behind Europe's dominant position in international relations during many centuries. In 1541,

King George VIII established the Royal Iron Works in Weald (Sussex) with a clear mission to develop the technology to produce iron cannons (Chaminade and Lundvall, 2019). Among the most noteworthy examples are the creation of ships with long-distance cannons or the development of steam power, which were decisive factors in increasing or reducing the British relative power in the world system.

The explanations offered by Kennedy for the rise and fall of the great powers in the international system base their arguments on the fact that "the relative strengths of the leading nations in world affairs never remain constant, principally because of the uneven rate of growth among different societies and of the technological and organizational breakthroughs which bring a greater advantage to one society than to another" (Kennedy, 1987).

During the 19th century, and after the Napoleonic era in Europe, the United Kingdom benefited from the consequences generated by the Industrial Revolution and from the impact these new technological developments had on the art of war. The British geopolitical hegemony during the 19th century is mostly due to the scientific and technological developments made on a large scale in the fields of transportation and communications throughout that century and to the British Empire's ability to adapt them to their naval and military prowess, which allowed it to develop a relative power greater to that of other empires, to maintain their control of international seas, and to conserve a hegemony over the rest of world powers.

In the 20th century, the link between scientific knowledge and international relations was intensified as a consequence of the acceleration of scientific and technological discovery and its larger impact on the dynamics of the international system. The advances ushered in by the so-called Second Industrial Revolution, which started toward the end of the 19th century, were enormous developments and innovations owing to the emergence of new energy sources (gas, oil, etc.), materials (copper, steel, etc.), transportation (planes and cars), and communication systems (telephone and radio), which immediately brought substantial changes to daily life. All these processes played an important role in global events at the beginning of the 20th century: the rise of Nazism, the creation of the Soviet Union, the fascist aspirations of Italy and Japan, and the development of the United States as a world superpower. One of the most outstanding examples was the great consequences of the Manhattan Project had to end World War II with the production of the first nuclear weapons.

After World War II, the role of STI grew even further as a result of two combined factors: on the one hand, the emergence of a new bipolar world order characterized by two superpowers (the United States and the Soviet

Union) that started an unprecedented arms race in the context of a nuclear threat, and, on the other hand, the emergence of the politics of *Big Science*, which marked the beginning of massive public investment programs for research and development (R&D).

Despite the criticisms garnered by the linear vision of scientific and technological development during the last quarter of the 20th century, the consensus on the necessity of making strong investments in STI has widely spread among the main actors of the international stage, as it has become a decisive element for socio-economic development. The relevance acquired by scientific knowledge in the international system and its proven impact on economic and social progress has led to many international actors (both public and private) to strongly invest in R&D with the purpose of developing their scientific and technological systems, stimulating their economy and narrowing the gap with the leading countries (UNESCO, 2015).

Throughout recent decades, multiple factors have allowed science to rapidly advance through new scientific and technological discoveries and inventions that open the doors not only to new knowledge but also to innovative applications. New scientific and technological areas include a wide variety of sectors and fields such as information technology, nanotechnology, biotechnology, cognitive science, and robotics. This means the emergence of new areas of science applications such as artificial intelligence, the Internet of Things (IoT), autonomous vehicles, 3D printing, or precision medicine, which promise to expand people's and humanity's possibilities in extraordinary ways.

The emergence of new scientific and technological areas in the international system means the possibility of discovering and producing new knowledge that would have applications and usages in a wide variety of areas which, at the same time, will inevitably open an unknown global scenery with regards to the consequences that these new technologies will generate. The key of these new scientific and technical fields is their capacity of becoming disruptive and innovative factors with a great impact on the entire world order.

Disciplinary approach

Despite the historical empirical relevance of scientific knowledge in the international system, in the academic field of international studies, the treatment of the phenomenon of scientific knowledge has been rather sporadic, reduced, and inarticulate. Whether because of its own internal epistemological and methodological confusion, the strong theoretical and pragmatic debates, or simply the interests it has served, the discipline of international studies has never given STI a privileged place in its analysis of the international reality.

Historically, within the international system, scientific knowledge has been understood as one of the conditioning factors, similar to geography, culture, or ideology. The most substantial contribution of STI to international relations has always been thought to be the contribution made to military development and its applications in the field of war. The discovery of new technical systems, tools, weapons, and innovations with military applications and which would represent an advantage in armed conflict has been considered the key factor in the link between STI and international relations.

This topic attracted special attention from researchers during the Cold War period when nuclear development and the threat of mass destruction became key topics in the global agenda. Later, as a consequence of the bipolar system, the appearance of an increasingly evident scientific and technological revolution, together with the rise of innovation as a key element of the capitalist economic system, some researchers started to study the impact of scientific knowledge on the economic development and the evolution of information and communication technologies (ICTs).

Traditionally, the discipline of international studies has primarily focused on topics such as security, war, political power, and diplomacy, but it has paid less attention to issues like STI. However, it could be said that this discipline has always had some sense of the importance of scientific and technological advancements because it does focus on STI but not explicitly. Instead, STI tends to be built into certain subject matters, including the domain of nuclear issues (e.g., Manhattan project, nuclear proliferation, and International Atomic Energy Agency (IAEA)).

Numerous internationalists (e.g., Waltz, Rosenau, Nye, Strange, Keohane, and Haass) from diverse theoretical perspectives approached the phenomenon of STI within the international system but always tied it to issues of security, economic networks, or communication systems. Certainly, the study of the role of STI and its impact on the main parameters and configuration of the international system was very limited; just a few researchers have been interested in the intersection between STI and international studies from a systemic point of view by reviewing the relevance and consequences of the scientific and technological revolution on the whole international system (Rosenau, 1990, 2003; Skolnikoff, 1993, 2002; Mayer, 2014, 2015; Weiss, 2005, 2015; Deudney, 2018).

This empirical and academic gap is very well described by many scholars. Despite the relevance of STI in the international system and the increasing number of researchers who are making efforts to study this new subfield of study, there is still a low number of scholars focusing on this issue within the discipline of international studies. As stated by Daniel Deudney (2018), the discipline "is still poorly equipped to theorize the sources and consequences of turbo change or to offer much in the way of actionable policy advice on this matter." In the same way, Charles Weiss (2015) pointed out, "the isolation of science and technology from the 'mainstream' of international relations like a curious anomaly in the 21st century."

What emerges from this brief analysis is the need for new research approaching the role of STI as a relevant phenomenon in the international system and with a high impact on the main processes, parameters, and transformations of the 21st century world order. Charles Weiss (2015) points out that the link between science and technology and international relations is still largely unexplored academically despite its relevance, and promulgates the need to advance scientific work and research in this regard: "Academic literature especially concerned with the impact of science and technology in international relations is relatively limited if we take into account the subject's importance."

The causes that explain the limited attention given by international studies to the holistic study of scientific knowledge as a relevant factor of the international system are multiple and none can singlehandedly explain this lack of systemic approach:

- The trend towards the super-specialization of modern science increasingly focuses on a small part of observable empirical phenomena, which distances scientific efforts from more global and holistic views.
- Scientific communities in international studies and STI have very different interests, theoretical and methodological frameworks, and methods that are not always easily reconcilable and have discouraged the search for intersection points, joint projects, and common disciplinary views.
- It is easy to recognize that topics linked to STI do not easily fit in the common theoretical framework of international studies.
- The existence of a historical bias in the field of international studies in favor of approaching topics preferably linked to security, military topics, and war is also evident, at the expense of other topics composing the global agenda.
- Another reason is the relative disciplinary immaturity of international studies, alongside a great epistemological and methodological confusion, that have stopped it from addressing empirical issues in a better way.

- Finally, the interests that many of the paradigms in the discipline of international studies have served have prevented more acute and profound approaches to such sensitive issues as the scientific-technological factor.

All these answers try to explain the limitations STI has had as a subject matter within international studies. Currently, the interest in STI is larger than in any other moment of history due to the key role it has acquired in global economic processes, in the resolution of issues of the global agenda, in the building of political power, in its innovative application in the military sector, or its core function in the virtual world. Because of this, it is possible to think that STI has become one of the main topics in the current international agenda and, at the same time, its study is necessary to understand the new international dynamics in the 21st century.

Nowadays, the study of STI from the perspective of international studies is especially interesting, as it is now seen as a key factor in the internal dynamics of the international system, which has created a greater awareness within the discipline of international studies of the need to study how STI affects the international system.

1.3 Rise of Science, Technology, and Innovation

The transition from a simplified bipolar order to a new, more complex, and multipolar system characterized by globalization, an incredible scientific and technological revolution, the coexistence of a large number and variety of actors, and the rise of a new global agenda have rendered STI a key factor in the 21st century's new world order. All these profound changes experienced by the world order in the past 30 years have created a new international context where scientific knowledge has acquired strategic value and a more relevant global positioning.

New role of scientific knowledge

In the changing international context of the end of the 20th century and the beginning of the 21st century, scientific knowledge acquires special and strategic relevance in the international agenda. As mentioned before, scientific knowledge has been an important factor in many societies throughout history, but now it has become essential for this new period of the international system because it is the main method for the generation of economic wealth, political power, military innovations, and social development.

Nowadays, economic and social growth is largely dependent on the strengthening of knowledge and human capital as the basic supplies of development. For that reason, there is a widespread consensus on wealth and a country's chances of development being highly linked to the consolidation of the sectors of STI, to the existence of a critical mass of scientists and professionals related to the productive sector, to the generation of research works that become scientific papers and patents, to the promotion of technological transference to the most dynamic economic sectors, and to the search of innovations aimed at solving structural issues.

In the current global context, scientific knowledge has reached a distinguished position in the international system due to its revaluation as a critical factor in the system, which transforms it into a truly valuable power resource for the majority of global actors. Eugene Skolnikoff (2002) argued that the interaction between science and technology and political and social factors "will be prime factors in determining the winners and losers among nations, the very economic viability of some, the forms of national economies and polities, the nature and cost of military conflict, the role and development of international organizations and the world's ability to meet the population, resource, food, health and environment issues we will face."

In this context, it cannot be strange, then, that we start talking about *knowledge economy* and *knowledge society* as concepts that define the new international reality.

Knowledge economy

The term *knowledge-based economy* or, more simply, *knowledge economy*[3] was popularized towards the end of the '90s to describe the transition from an economy based on production and industry to one mainly based on knowledge. As understood by Foray (2006), we can generally consider the knowledge economy to be based on the usage of ideas more than physical skills, or in the application of technology rather than the transformation of raw materials or the exploitation of the workforce. Natural and industrial resources have lost a large part of their ability to explain the disparity of productivity and growth between countries, and the economy's center of gravity has shifted to the production of goods and services with information content, to technological innovations, and to the creation and accelerated

---

[3] The concept of *knowledge economy* was popularized with Peter Drucker's book *Age of Discontinuity* (1969), where he attributes the term to the economist Fritz Machlup and where the idea of *scientific management*, previously developed by Frederick Winslow Taylor, first appears. Ducker uses the term to describe the differences between manual workers and knowledge workers: while the former work with their hands and produce goods and services, the latter work with their minds and produce ideas, information, and knowledge.

productive use of new knowledge. Castells (1996, 2005) defines the knowledge economy as "an economy in which the increase of productivity does not depend on the quantitative increase in the production factors (capital, work, natural resources), but on the application of knowledge and information to the management, production, and distribution, both in the processes and the products."

The key to the emergence of a knowledge economy has been the newly acquired relevance of the economic competitiveness between the different actors of the international systems (and not only companies) and the role played by scientific knowledge in that competition. The current international economic system demands competitiveness, as well as an increase in commercial exchange and investment, to ensure prosperity and economic growth. Nowadays, the only competitive edge comes from knowledge and innovation (Porter, 1998, 2011), where the ability to innovate is the cornerstone of competitiveness and the enterprise becomes the mediator that transforms *inventions* (understood as the product of new knowledge) into *innovations* (understood as the economic valuation of the new knowledge) (UNESCO, 2005). The crucial dynamics of the system is scientific knowledge, assumed by companies and countries as a decidedly endogenous factor of growth, which is translated into applied technologies (management, processes, and materials) through the sequence *research & development & innovation (R&D&I)*. As explained by Porter (1998, 2011), knowledge is being developed and applied in new ways, the lifespan of products is shorter, and the need for innovation and competitiveness is growing.

Following Castells's (2000, 2009) analysis, three characteristics define this new knowledge economy.

(i) It is an economic system that considers knowledge the basis of production, productivity, and competitiveness for companies, regions, cities, and countries.
(ii) The predominant economic activities are globally articulated and function as a single unit in real time around two economic globalization systems: the globalization of interconnected finance markets and the organization on a global level of the production and management of goods and services.
(iii) It is seen as an economy that works in networks: within the company, between companies, and between companies and other actors.

The impact this economic phenomenon has on the international system is revolutionary. Currently, the successful economic performance of the main international actors is increasingly determined by their ability to generate, acquire, and utilize knowledge. As Foray (2006) points out, this new trend is also characterized by the great speed in which scientific knowledge is created and accumulated and by the substantial decrease in the cost of codifying, transmitting, and acquiring it.

Knowledge society

The particular role being played by scientific knowledge in the current phase of the international system has gone beyond the strictly economic sphere into the rest of society, generating an extensive and powerful impact in other domains. In this context, it seems appropriate to call this phenomenon *knowledge society*, as it reflects, on the one hand, the economic significance STI has on the globalized world and, on the other hand, it includes the social, cultural, and political relevance also acquired by scientific knowledge.

The knowledge society has burst in as a global phenomenon that includes and transcends other emerging events in the international system, such as the *knowledge economy* or the *information society*.[4] Essentially, the concept of the knowledge society refers to a new type of society characterized by the intensive application of knowledge in all walks of life and where knowledge becomes the main source of production, wealth, and power.

UNESCO (2005) has adopted the term *knowledge society*, or its variant *knowledge-based societies*, within its institutional policies with the aim of incorporating a more comprehensive and pluralistic approach to the phenomenon, including its social, cultural, educational, and political transformations. In the same vein, Innerarity (2011, 2013) holds that limiting the impact of scientific knowledge to a mere economic element does not do justice to the relevance it currently has for society as a whole.

The rise of the knowledge society is a global process that will not completely replace the old industrial society but will coexist in a spectrum of different situations over a long transition period. Everything suggests that the road to a true knowledge society will be a slow, complex, heterogeneous, and conflictive process that will largely depend on actors' capacity to extend the benefits of the new age of knowledge to the entire international society.

---

[4] The phenomenon of the *information society* refers to a society where the creation, manipulation, distribution, usage, and integration of information are economically, politically, and culturally central activities. Implicit in this kind of society is the idea of a growing technological capability to store an increasing amount of information and circulate it more quickly and with greater diffusion capacity.

1.4 International Scientific Relations

The evolution and prominence of STI throughout the history of international relations and the rise, in recent decades, of a new world order where scientific knowledge plays a much more significant role, have resulted in the appearance of a new global context where the links between international affairs and scientific and technological development have notably intensified. The confirmation of this new reality is what allows us to recognize the emergence of a new subfield of empirical and theoretical analysis that we call *international scientific relations (ISR)*.

New empirical phenomenon

ISR must be understood as an empirical space focused on the links between the development of scientific knowledge and its applications in the fields of STI with actors, phenomena, parameters, and changes in the international system. The changes that have taken place in the world order in recent decades have meant a reevaluation of STI as a tool for economic, social, and political development, which meant, at the same time, an increase in the actors, relations, and processes linked to scientific knowledge, which are relevant and have an impact on the new world order. The dynamics of international relations is changing fast, and topics such as STI are acquiring an increasingly prominent role in the 21st century's global topic agenda.

The clearest evidence of this new empirical phenomenon is the visible consequences STI have on the current international reality: from discussions at the highest political and intergovernmental level concerning the application of scientific advances in topics such as the environment or energetic development, through new threats such as cyberattacks or the spread of global pandemics, to the appearance of new emerging technologies that are changing each and every aspect of our daily life.

In recent years, the acceleration and expansion of the technological revolution has rapidly changed the landscape of international relations. New scientific knowledge, emerging technologies, and an impressive number of outstanding innovations and developments have increased the complexity of the international system. The use of drones and artificial intelligence (AI) applied to military operations, hackers trying to manipulate presidential elections, advanced robotics changing business models, a growing globalized virtual society, and new public policies to promote the digitization of manufacturing are some of the newest and most influential factors in the current global agenda. At the beginning of the 21st century, STI is an omnipresent participant in the narratives of world politics and is becoming a strategic source of economic development, political influence, military power, and social innovation.

New subdisciplinary field

In the context of these extraordinary changes that are taking place in the international system, the search for explanations that address the new reality represented by STI in the current international system must be seen as a social and academic need. The major restructuring suffered by the international system and the magnitude and speed of the changes we are witnessing make unavoidable the double task of advancing in the analysis and understanding of the global changes we are seeing as well as rethinking the theoretical and methodological framework we use to give concrete answers on the phenomena examined.

The discipline of international studies has addressed some of these issues, but it is still a pending task to build analytical frameworks that better explain the changes we observe at the empirical level as a result of the action of STI in the international system. This empirical field must be understood as a potential area of theoretical study through a subdiscipline in international studies, namely ISR.

The main objective of the subdiscipline of ISR is the study, within the international system, of all the connections and interactions established between STI and international relations. This is a relationship of mutual influence between both spheres: on the one hand, the production of scientific knowledge and its applications in STI has a direct impact on international economy and politics, as shown by the technological development applied to the military–industrial complex or the reevaluation of scientific knowledge as a strategic resource of the new international economy; on the other hand, the global dynamics of international relations has a decided influence on scientific and technological development, as shown by the geopolitical and geoeconomic impact that the growing R&D investment of countries such as China has.

The intersection between STI and international affairs opens a new and extended research agenda within international studies with topics such as STI diplomacy, multilevel governance of knowledge, international cooperation networks, the impact of emerging technologies, or the knowledge divide. The relevance of this new area or subfield of study—*international scientific relations*—has increased in recent years and is becoming an essential point of interest to fully understand the changes in the 21st century world order.

Although this new subdiscipline must concentrate itself on the scientific study of a very reduced part of the international reality, due to the nature of its subject matter, ISR is required to create a dialogue with other subdisciplines within international studies, with other disciplines within social sciences, and with other technical disciplines of the STI area in order to carry out a multi- and interdisciplinary approach to study the empirical phenomena. At the same time, it is expected that the development and evolution of this subdiscipline will serve for the design of new theoretical schemes of interpretation and, at the same time, that it will have practical utility for the conformation of public policies and private strategies linked to STI in the international system.

Challenges for international scientific relations

STI and international relations have established a deep and growing relationship within the frame of the new international system. This has meant the emergence of ISR as a specific field of study where empirical realities and phenomena linking STI with international affairs are observed and addressed.

This promising subfield of study faces several challenges that must be tackled as soon as possible if it wants to reach truly valid scientific explanations of the phenomena observed in the current international system. We must specially consider the following challenges:

- The difficulty of studying empirical phenomena that show great dynamism and variability.
- An empirical phenomenon that can greatly impact the main systemic parameters (actors, interactions, processes, configuration, etc.).
- The lack of theoretical and practical precedents within the discipline of international studies that holistically address the interaction between STI and international affairs.
- The necessity of creating new theoretical frameworks and adapting old focuses to give an account of the new reality of STI in the international system.
- The urgency in building methodological schemes that include new techniques and instruments for data collection and data analysis.
- Advancing in the task of promoting dialogue not only with other disciplines in social sciences but also with other scientific and technological fields to reach a more complete and holistic analysis through more interdisciplinary approaches.

There is no doubt that this subdiscipline has a great challenge ahead when addressing the phenomenon of scientific knowledge and discerning its role in the international system. Despite the existence of a few projects on the topic, STI is still generally treated as an isolated, secondary element with a limited impact on the international system. This confirms the assessments of experts such as Weiss, Mayer, or Deudney, who point out the need for a greater disciplinary development on the subject.

It is evident that the challenges faced by the subdiscipline of ISR are multiple and complex; however, the need for answers and explanations of the influence and impact of the new empirical phenomena observed in the international system requires innovative efforts to face it. The present research aims to be a step in that direction.

Part 2

ANALYTICAL FRAMEWORK

Chapter 2

INTERNATIONAL CONTEXT

In recent decades of the 20th century, the world started a frenzied transformation, prompted by multiple global processes that have operated in parallel within the world order and have generated significant changes in the whole of the international system: the end of the Cold War and the Soviet Union, the new trends towards globalization and regionalization, the new challenges to the Nation-State, the evolution towards a new stage of capitalism, and an unprecedented scientific and technological revolution are some of the more relevant processes that have affected the international system and that are creating a new world order in the 21st century.[1]

This new context implies substantial changes in the international order, that have generated (and still are generating) a structural transformation of the political, economic, social, technological, educational, and cultural spheres. The majority of experts agree that nowadays we are experiencing a period of systemic transition within international relations, characterized mainly by the speed, magnitude, and complexity of the changes.

To perform a detailed, in-depth analysis of this new global order, it is necessary to study the main parameters of the international system (actors, interactions, and processes) to understand how these elements would shape a new world order.

2.1 *Actors of the International System*

The first of the elements we must consider are the main *actors* that take part in the global dynamics. In our contemporary international reality, there coexists a large range of actors with varied characteristics according to their number, power, role, and interests. The post–Cold War international system has opened a new stage where the traditional actors of international relations

---

[1] By world order we refer to the combination of parameters within which the relations between the actors of the international system are sustained, seeking to satisfy expectations and realize aspirations (Dallanegra Pedraza, 2010).

coexist alongside new actors and where the Nation-State is forced to interact with non-State actors that exhibit a renewed prominence.

Changes in the role of the Nation-State

The world order created by the Peace of Westphalia in 1648 crowned the Nation-State as the main actor of international relations and made it the main actor responsible for security, administration, cooperation, justice, and social integration. However, in recent decades, the State is being challenged by new actors and non-State entities, such as transnational corporations (TNCs), intergovernmental organizations (IGOs), or nongovernmental organizations (NGOs).

Celestino del Arenal Moyua (2009) points out this new reality where the Nation "has been forced to share its international prominence with other international actors and, as such, is no longer the only or main actor in shaping the structure and dynamics of the international society," and, in parallel, identifies the advance of new non-State actors whose "spectacular growth and prominence is a direct consequence of the interdependence, globalization, and transnationalization dynamics." As Pierre Calame (2014) reflects, it is not about pretending that States have become unimportant entities on the international stage, but on the contrary, they are still central actors in military matters, international agreements, and public policy planning and management; however, their undisputed role has been challenged by the appearance and development of new actors that are trying to reduce their power.

New actors

In the last decades, new actors and non-State entities have gained prominence and starting to challenge the traditional hegemony of the Nation-State. Since the '70s, authors such as Robert Keohane and Joseph Nye have pointed out the emergence of new entities, even though it was not until after the end of the Cold War that they became more visible.[2] The main non-State actors that have emerged in recent decades on the international stage include the following: TNCs; NGOs; IGOs; regional

---

[2] These non-governmental transnational actors in world politics were defined by R. Keohane and J. Nye as "a set of transnational activities surrounding numerous kinds of border-crossing contacts, coalitions and interactions that are not controlled by organs of government" (Keohane and Nye, 1987, 1998).

processes; individuals, by which we mean actors on an individual or collective level (i.e., public opinion, the budding global civil society); religious, ethnic, cultural, and social movements; nationalist and/or separatist movements; and international mafias, drug cartels, smuggling networks, hackers (organized transnational crime), and terrorist groups. Each one defends its own interests and influence, interacting on a more complex and interdependent stage.

States still have a privileged role in the international system, but we would be mistaken to consider these non-State actors as secondary entities as far as their size, capacity, or influence is concerned. On the contrary, these new actors, by virtue of their vocation, dimension, flexibility, and organization, are challenging the role of the traditional Nation-State.

Multipolar world

The coexistence of new and old actors in the international system has created what James Rosenau (2003) called the *bifurcated global structure*, which refers to the idea of the coexistence, in terms of power and decision, of State actors and non-State entities and, as a result, creation of a global configuration composed of a vast network of international actors with very different levels of power and interests. Susan Strange (1996) argues that "power has shifted horizontally, from States to markets, and consequently to non-State authorities whose power comes from their market quotas." This means the emergence of a new international configuration with overlapping authorities where governing is much more difficult. To identify and characterize this type of overlapping powers, Hedley Bull (1966) coined the concept of *neomedievalism* (Friedrichs, 2001), which Strange also uses to explain why the 21st century's international system is much more similar to the world of the 12th century rather than the 18th or 19th centuries with regards to the structure of authority. Thomas Risse (2008) summarizes this process pointing out that "it is certainly true that transnational actors–from multinational corporations (MNCs) to Non-Governmental Organizations (NGOs)–have left their mark on the international system and that we cannot even start theorizing about the contemporary world system without taking their influence into account."

What is evident from this analysis is that the emergence of new actors has given the international system a new configuration characterized as *post international* and *polycentric*. Alongside governments, there are a growing number of increasingly powerful transnational actors involved in international dynamics and processes.

2.2 Interactions and Relations within the International System

The second element to analyze in our present international context consists of the *interactions* and *relations* generated between different international actors. The acceleration of phenomena such as globalization, the scientific and technological revolution, and the existence of more global actors taking part in the international system are some of the main factors that are shaping new types and a greater number of linkages among actors.

Interdependence

The traditional interactions and relations between states (mainly bilateral) have given way to a new international scenery with a larger quantity of actors (both State and non-State) that make a greater number of linkages (e.g., multilateral, regional, international, transnational). Since the end of the '70s, Robert Keohane and Joseph Nye have described the emergence of a new international system precisely characterized by the existence of new non-State actors interested in global affairs and by the acceleration and intensification of the contacts among them, and this was generating a new context of *complex interdependence*.[3] In essence, Keohane and Nye pointed out the existence of new and more complex linkages between the system's actors, at least in some regions (developed countries) and in certain spheres (economy, communication, etc.).

After the Cold War, *complex interdependence* accelerated and intensified among all the actors of the international system because of the special conjunction of three simultaneous global phenomena: first, the existence of an increasingly globalized world, which means an increase in the interconnection and interrelation of the majority of actors, at least economically, financially, and communicationally; second, the surprising scientific and technological revolution, which generated an incredible advancement in communications and transportation that has facilitated and encouraged relations and links among all areas of the planet; and third, the variety and intensity of the interactions have increased as the number of State and non-State actors, with interests and the capacity of influencing the international system, has grown.

---

[3] *Complex interdependence* is built on three combined characteristics: (i) the existence of multiple channels, which allow for a larger and better connection between State and non-State actors; (ii) the existence of a new global agenda defining the relations between actors; and (iii) the consideration that military resources are not the only power resources in the international system (Keohane and Nye, 1987, 1998).

New global agenda

At the same time, the interactions between these actors were intensifying, with the number of topics that motivated those links also being diversified. Although during the majority of the Cold War period, the topics related to military security caught the interest of international actors, since the end of the '70s, the existence of a *complex interdependence* between the different actors, which forced them to discuss a more global, numerous, and nonhierarchical global agenda, has drawn increasingly more attention. The end of the bipolar order allowed new topics that had been displaced or subordinated due to reasons of security or defense to reappear, which allowed the establishment of a much more diversified, extensive, and interdependent global agenda. This new historical period shows us an agenda where some topics have emerged, some have strengthened, and some have changed:

- New topics are appearing due to the historical context such as terrorism, cyberattacks, drug trafficking, global corruption, pandemics, humanitarian crises, and the proliferation of weapons of mass destruction that transcend national dimensions and become relevant global challenges.
- Among the topics that have increased their relevance in the international agenda, we find environmental issues, poverty, inequality, migrations, human rights, global health, the economic and financial situation, among others.
- Lastly, among the topics that have changed are those linked to strategic defense and military security matters. This does not mean, however, that actors have lost interest in such important international affairs issues, but that nowadays that interest has moved on to new ways to understand power and defense.

The Agenda 2030 set in 2015 by the United Nations General Assembly is a good example of the new issues in the international system. The Sustainable Development Goals (SDGs) or Global Goals are a collection of 17 interlinked goals that were set in 2015 by the United Nations General Assembly and are intended to be achieved by the year 2030.[4]

---

[4] The 17 SDGs are (1) No Poverty, (2) Zero Hunger, (3) Good Health and Well-being, (4) Quality Education, (5) Gender Equality, (6) Clean Water and Sanitation, (7) Affordable and Clean Energy, (8) Decent Work and Economic Growth, (9) Industry, Innovation, and Infrastructure, (10) Reducing Inequality, (11) Sustainable Cities and Communities, (12) Responsible Consumption and Production, (13) Climate Action, (14) Life Below Water, (15) Life On Land, (16) Peace, Justice, and Strong Institutions, and (17) Partnerships for the Goals.

Types of linkages

The new international agenda consists of numerous issues that are tackled by international actors through a large variety of interactions. We can differentiate and classify those links into four major categories:

(i) *Conflictive links* have been one of the most studied interactions in the field of international studies. The traditional paradigms (realism and neorealism) understand the international order as a naturally anarchic system where States, as main actors, establish conflict relations among themselves with the aim of imposing their own *interests* through the use of coercion. Although this vision has been reconceptualized in recent decades (globalists, international political economy, critical theory), allowing and recognizing the existence of other variables, actors, and linkages, still this paradigm believes in the use of power by a majority of the actors in the absence of a global government or supranational organization (a *Leviathan*) that is able to regulate the relations between its members. In this sense, conflictive links seem to be a constant in the international system, even though the nature and the shape of the conflict vary in each historical moment.

(ii) *Cooperation dynamics* appear in the field of international studies as the other side of conflict relations and can be understood as traditional links in international relations. Since the beginning of the 20th century, these cooperation relations are studied in the field of the international system (idealism) where the main actors establish links of harmony and articulation of interests because the objectives they pursue can be considered complementary rather than antagonistic. Even though conflict cannot be avoided, it is possible to transform the international system into a lasting peaceful scenario, or, how Immanuel Kant called it, *a Perpetual Peace*, through the establishment of mechanisms as international law or the creation of intergovernmental organizations. Since the mid-70s, the globalist, led by Keohane and Nye, went a little further in the concept of cooperation, considering the existence of an international pluralism mandating the cooperative interaction between national interest groups in transnational structures. The existence of interdependencies and the creation of mutual influence tools between international actors have allowed the number of cooperative agreements and *international regimes* to multiply (Keohane and Nye, 1987, 1998), which has led to substantial agreements in strategic topics such as armament control, the navigability of oceans, or the regulation of commercial exchanges.

(iii) *Competitive interactions* have become one of the more relevant and frequent dynamics in the international system. Practically all paradigms in international studies assume the existence of competitive relations as a way to explain the power dynamics between different actors of the international system in the pursuit of imposing their own particular interests. In the last few decades, the traditional international logic of military competitiveness linked to security affairs has given way to new forms of competition for economic resources, which have also become strategic elements for global actors. In our current international context, competitive logic has become a common practice of actors with the aim of obtaining advantages over their peers in issues as relevant as the military, the economy, or technology.

(iv) This type of linkage is characterized by the articulation of cooperative interactions with others of a more conflictive nature, giving way to relational processes that are deeply rooted in international structures, such as *influence, hegemony, domination, dependency, or even exploitation*. Most paradigms within international studies understand these links as a habitual mechanism within the international concert of domination of a hegemon, or of several hegemonic states, over other less powerful states. However, from more critical trends (Marxism, postmodernism, and critical theory), authors such as Robert Cox, based on Antonio Gramsci's categorization, consider it more pertinent to speak of a *hegemonic order*, where "*coercion*" and "*hegemony*" are sources of power and control in the international system.[5] What characterized the international reality during the Cold War was the hegemony of two superpowers over their own blocks of influence, where they projected their power through a combination of persuasive and coactive actions, aiming to impose and maintain their particular interests as general interests of the international system as a whole. The end of the Cold War broke this model and maintained, for a time, the illusion of the *Pax Americana,* which came crumbling down with the 9/11 terrorist attacks.

The interactions and relations among actors within the international system have accelerated and intensified in their number, type, and matter. The rise in the number of interactions among actors has increased the complexity of these links and, at the same time, stimulated the emergence of new processes and phenomena on a global level.

---

[5] R. Cox defines hegemony as "a structure of values and understandings about the nature of order that permeates a whole system of states and non-state entities" (Cox, 1993).

2.3 Main International Processes

From the point of view of systemic analysis, it is essential to study a third element linked to the main *processes* generated within the international system itself. These processes must be understood as internal and essential mechanisms and movements that make the system work and, in international relations, are identified with the main international practices established in the world order, influencing the system as a whole. In the framework of our current international stage, we can recognize some fundamental political, economic, social, and scientific and technological processes within the international system that affect the international dynamics and are making profound changes to the global system of the 21st century.

End of the Cold War

The first of these processes is the end of the Cold War. The end of this conflict has meant the extinction of the international order that reigned from the end of World War II until the definitive dissolution of the Communist Bloc in 1991. During this whole period, the international system was dominated by the ideological clash between two non-European superpowers (the United States and the Soviet Union) championing an antagonistic economic and political system and maintaining an indirect confrontation all across the world map. The end of this cycle was marked by the historical sequence that started with Mikhail Gorbachev's 1985 projects for economic and political change in the Soviet Union, known as the *Perestroika* (economic reform) and the *Glasnost* (political reform), the fall of the Berlin Wall in November 1989, and the coup and later dismemberment of the Soviet Union in August 1991.

Following Fred Halliday's (2000) analysis, we find three main factors that explain the end of the Cold War: first, the Soviet Union's failure to compete with the United States in the new dimensions offered by the scientific and technological revolution, fundamentally in the development of new technology; second, the Soviet bloc's failure to foster high rates of economic growth in a controlled economic system; and last, the explicit acknowledgment from the Soviet elite that the United States was more developed not only in strategic-military matters but also in economy, especially in the production of consumer goods and services. The Soviet authorities' awareness of this conjunction of factors was what led to the reform process known as the *perestroika* and later to the subsequent triggering of historical events that led to the end of the Soviet bloc.

The core in this process is the magnitude of the change caused by the end of the Cold War and the impact it had on the transition to a new ordering of the international system. The fall of the bipolar structure that identified the period of the Cold War as a whole implied, at least, three main consequences over the configuration of the post–Cold War international system:

(i) The collapse of one of the superpowers (the Soviet Union) left the role of the main actor in the international system to a single political and economic reference, that is, the liberal-capitalist democracy of the United States, but it also meant the emergence of new poles with the goal of balancing the hegemon's power.
(ii) The end of an international agenda marked by military issues, which gives way to the treatment of new topics, previously subordinated to defensive logic.
(iii) The end of the bipolar strategic alignments and the actors' disciplined subordination to one of the antagonistic blocs.

These consequences, clearly distinguishable in our current global context, have started to have repercussions in the shaping of a new world order. As pointed out by Mikhail Gorbachev, one of the lead actors of that process, the changes in the world after the end of the Cold War, and the difficulties for the emergence of a new political and economic order must be understood as manifestations of much more profound changes in the development of humanity.

Challenges to the Nation-States

A relevant second process generated in the international system is the challenge to the Nation-State's traditional role in the international political system. It is indeed two phenomena linked to the same process: *The crisis revealed in the Westphalian order and the changes in the international political order.*

The first process is the crisis revealed in the Westphalian order that can be explained by a confluence of internal and external factors. On the one hand, the State must face a double external pressure due to (i) a new global context characterized by the growth of the Nation-State's frontiers and (ii) the increase of local and subnational identities since the end of the Cold War. On the other hand, the traditional territorial and political–military power, which was the exclusive monopoly of the Nation-State, is being challenged by informal players that are encouraged by the resources with which the State must be constituted, such as a large part of the economic regulatory functions that are now transferred to international organizations and even to private actors[6] (Smouts, 2001).

---

[6] According to Habermas, there are three main aspects of the erosion of the Nation-State's prerogatives: (i) the decline in control capacity, (ii) the growing deficits in the legitimacy of

A second process is the modification of the international political order. The so-called Westphalian order has been the system of political organization based on the principles of territoriality and the sovereignty of the Nation-State for the past 350 years. This conceptual model has allowed the modern State to transform into the lead actor in the political-administrative organization of international society, exercising functions of security, administration, regulation, cooperation, and justice. However, the changes in the global system that have started to occur in the last few decades, and especially the strong transnational and globalizing dynamics, have raised questions as to the continuity of the Westphalian order, which has its basis in the absolute sovereignty of the modern State over the rest of the actors of the international system.

In any case, the Nation-State, for the first time in hundreds of years, finds its position as the lead actor in the international system threatened, which could imply a change in the international political order. For some, such as Strange (1996), it is the end of the system of Westphalian States, which she renames as *the Westfailure system*, and the permanent withdrawal of the State, whose power will shift over to other non-State authorities (The Defective State). For others (Halliday, 2000; Halliday et al., 2006), however, the Westphalian system will not disappear but will take on new meanings and Nation-States will go on conserving considerable power quotas because of the change to State coalitions, formal (EU) and/or informal (BRIC, G7, G20, etc.) must not be confused with the permanent dissolution of states.

Scientific and technological revolution

The third process that has taken place in the international system is the one known as the *scientific and technological revolution*. It is a profound change that affects the very foundations of the international system and, together with an unparalleled scientific and technological development, is modifying each and every aspect of day-to-day life and is decisively impacting political, economic, social, and cultural activities. The expressions *scientific and technological revolution*, *technological revolution*, *scientific and technical revolution*, and *digital revolution* are concepts used to refer to the same phenomenon: the technical transformations (and their social and economic implications) that have taken place in the international system from the last quarter of the 20th century to present day. The relevance and impact of this scientific and technological revolution has

---

the decision-making processes, and (iii) a growing inability to carry on the organizational and governmental functions that help assure legitimacy (Habermas, 2000).

led to the frequent usage of expressions such as the *Third Industrial Revolution* and *Fourth Industrial Revolution* to compare them with the first two great transformations in human history that deserve the name of revolutions: the Neolithic Revolution and the Industrial Revolutions.

The main precedents for this truly global scientific and technological revolution go back to the end of the 1970s, when the techno-productive paradigm, upon which the capitalist world had based its growth for almost three decades, went into crisis. Thanks to this revolution, media and transport strengthened links and relations between all corners of the planet, which has brought humanity the closest it has been to the concept of *global village*.[7]

The main characteristic of this scientific and technological revolution is the change from analog mechanisms and electronic technology to the digital technology developed since the 1980s, owing to the massive production and generalized use of digital circuits and their technological derivatives, such as computers and mobile phones. The key element in this revolution is the microprocessor, as it exhibits constant increases in performance and allows technology to be attached to a wide range of items. Equally important has been the development of information and data transfer technology, such as the computer network and digital and internet broadcasting. Smartphones have had an exponential social penetration and play a very important role in the digital revolution, simultaneously providing instantaneous communication and connectivity.

The speed and depth of the intensive application of new technology to many facets of daily life makes it hard to evaluate the impact all these changes are generating in society. According to Bauman (2000), Castells (2005), and Schwab (2016), it is possible to describe a few of the more visible transformations:

- The enormous range of research and development possibilities in new areas and branches of knowledge, unimaginable until now.
- The emergence of new industries and new goods and services sectors.
- More flexible organizational structures and production systems.
- The dematerialization of manufacturing and commerce.
- The transformation of companies, no matter the sector, into information processors.
- The existence of global markets that are simultaneously segmented and global.

---

[7] The concept *global village* was first coined in 1968 by Canadian philosopher Marshall McLuhan in his book "War and Peace in the Global Village" as an expression of the growing human interconnectivity generated by electronic communication.

- The elimination of hurdles for cross-border circulation of goods, services, and capital.
- Changes in time and space references that are totally discarded in the traditional way.

Now in the second decade of the 21st century, and as a consequence of the magnitude, speed, impact, and projection of the scientific and technological revolution we have experienced for the past few years, some experts point out the advent of a *Fourth Industrial Revolution*, different to the *Third Industrial Revolution*. Klaus Schwab (2016, 2018) points out that there is clear proof to think that we are at the beginning of a Fourth Industrial Revolution, starting in the 21st century, building on the framework of the previous digital revolution. Essentially, "the speed of current breakthroughs has no historical precedent. When compared with previous industrial revolutions, the Fourth is evolving at an exponential rather than a linear pace. Moreover, it is disrupting almost every industry in every country. And the breadth and depth of these changes herald the transformation of entire systems of production, management, and governance" (Schwab, 2016). Following Schwab's arguments, what makes this new revolution different from the previous ones is the existence of new scientific and technological areas (such as artificial intelligence, nanotechnology, or renewable power) that mix and interact, at the same time, in the physical, biological, and digital areas, opening a wide and previously unthinkable array of possible applications.

Changes in the capitalist economic system

The fourth process to be considered within the international system is the *changes in the capitalist economic system*. Ever since the first signs of budding capitalism in 15th-century Amsterdam or its better-documented beginnings in 16th-century England up to the most evolved stages of the 20th-century global financial capitalism, the changes in the history of capitalism have always been one of the most impactful processes for the international system as a whole.

The long evolution and mutation of capitalism as society's system of economic organization have had, during a large part of the 20th century, their hegemony openly challenged by antagonistic ideologies such as fascism or communism, which directly clashed with capitalist logic and advocated for replacing or overcoming it. However, the defeat of the fascist regimes in World War II and the fall of the Communist Bloc after the Cold War meant the almost exclusive hegemony of capitalism as the regulating force behind global economic activities. This supremacy has only been possible due to the fact that, through all of its historic evolution as an economic model, the capitalist system has mutated, transformed, and reconfigured itself numerous times to be able to stay in power.

One of the most transcendental changes happened at the start of the '70s, between 1968 and 1975, during a period where the economic features established in Bretton Woods (understood as the economic architecture created for the post–World War II period) were modified towards a subordination of domestic economies to the exigencies of a global economy. States started to internally justify some accountability with the outer world through the usage of terms such as globalization, interdependency, and competitiveness (Cox, 1993, 1996). From there, the formation of an international economy, characterized by the globalization of production and finances thanks to the usage of new technologies, became increasingly evident: production became global, which allowed many companies to use their territorial divisions to maximize factor cost reductions; at the same time, finance began to develop a path of decoupling from production to become an independent and autocratic power over the real economy.

These changes in the economic system started a new stage of capitalism, which became an essentially *postindustrial* and *post-Ford* economy. This meant the change from a production system based on an industry ruled by companies, syndicates, and State regulation to the production of management without industry (in terms of territorial location) and without regulation, formed by the actions of companies and the competition between them. Economy is increasingly outsourced, centered in the intensive usage of intellectual capital, increasingly globalized, and much more flexible and mobile due to the usage of new information and communication technologies (ICTs). At the same time, the process of global economic integration has benefited from the reduction of the time and cost of transport and communications, as well as the breaking down of barriers for the international circulation of goods, services, capitals, knowledge, and even people and workforce, with some limitations (Stiglitz, 2002). A consequence is the emergence of a new form of social and economic organization, based on the volume, speed, and ubiquity of the generation of scientific knowledge and its immediate application for technological change.

Globalization

The analysis of our current international context could not be understood without seeing *globalization* as one of the most recognized phenomena of our current global context. This is a phenomenon of multicausal origin, of enormous proportions, and with unforeseeable consequences for the entire international system.

During the first half of the '90s, and despite the imprecision inherent in the new world context that was beginning to take shape, a consensus that globalization was a key factor in the changes the international system was going through was slowly emerging. Essentially, "globalization connotes the stretching and intensification of social, economic, and political relations across regions and continents. It is a multidimensional phenomenon that embraces many different processes and operates on many different time scales" (David Held, 2000). A more interconnected and interrelated world, at least in some areas (economically, financially, and communicationally speaking) and in some specific geographic coordinates (developed countries). As Richard Haass (2020) points out, "increasing global interconnection – growing cross-border flows of people, goods, energy, emails, television and radio signals, data, drugs, terrorists, weapons, carbon dioxide, food, dollars, and, of course, viruses (both biological or software) – has been a defining feature of the modern world."

Speaking of globalization implies, firstly, a scientific and technological development on a global level, which allows action beyond the traditional referents of time and space. On the one hand, with the overcoming of time and *instantaneousness*, international society enters a new dimension, global time, which entails the virtual death of time as an irreversible temporal separation and a physical obstacle to communication. On the other hand, the overcoming of space and *ubiquity* implies the possibility that actors can virtually appear in different places at the same time, thus going beyond the traditional concept of space as something inextricably linked to the territory. The *network society* allows us to connect an increasingly larger number of international actors and diverse events, previously unconnected due to geographic or temporal distances or cognitive barriers; this means, in short, a contraction of the space-time dimension.

Generally speaking, globalization must not only be understood as a product of the end of the Cold War, but rather as an evolutionary process with multiple historic precedents. In reality, since the 1970s, if not since 1840, *global interdependence* has been a widely debated topic. Some authors even consider that the historic novelty of this globalizing process is exaggerated, because States were exchanging larger percentages of their GDP before World War I than they are doing now, and people have emigrated in larger quantities than they do now (Hirst and Thompson, 2002). Whatever the case, it was in the '70s when the increase of commerce and other links between countries was seen as proof of the growth of interdependence; in the '80s and '90s, it was clear that something more important than the mere interconnectivity between actors was taking place. While global interdependence may have been a feature for centuries, it is the broadening and deepening of this phenomenon that has led to globalization. Modern interdependence has no precedents in its features, and currently, it is not only linked in economic markets that are being formed but also between individuals, cultures, and countries.

Despite the proven strength of the globalizing process, we must also consider the phenomenon's unequal reach in the economic and political spheres. We are currently living in a highly globalized, interconnected, and interrelated world in terms of economy, finances, and communications, but there still is no clear effect of globalization on the political sphere. Economic globalization is more evident with the growing integration of national economies into an international economy through commerce, foreign direct investment, the flow of capital, the international movement of workers and people, and the development of technology. In contrast, political globalization seems like a more distant and less palpable phenomenon when observing the international reality. Global interconnectivity has not created an institutionalized political framework with global goals, similar to what we could consider a global government, thus maintaining the old inherited institutional structure of the old bipolar order.

Currently, the international political architecture shows a group of intergovernmental institutions, created after the end of World War II, that desperately seek to redefine their main functions in light of the changes that have taken place, but still unable to show themselves as global agents capable of joining global topics and actors under their regime. Despite that, the risks and threats on a global scale, such as terrorism, environmental degradation, global pandemic, or economic crisis, have reinstated the debate on the need for some form of global political governance to face those challenges.

The concept of globalization is also associated with the confusion generated by the coexistence of contradictory megatrends within the same globalizing phenomenon. On the one hand, the trend of economies towards globalization, the global governance of world politics, and the homogenization and secularization of the cultural value systems; on the other hand, and at the same time, the trend of fragmentation, which can be seen in the resurgence of nationalism and populism, the new rise of religious fundamentalism, ethnic protectionism, and cultural relativism (Rodrik, 2011, 2020).

A phenomenon as complex as globalization has generated a strong debate between proponents and detractors who discuss its scope and consequences. On the one hand, globalization optimists (identified by a liberal position) point out that, in its basic dynamics, the world economy would become internationalized and would be dominated by uncontrollable market forces, which is seen as a positive fact, since this would result in the emergence of a global economic system of free world trade, which would improve the welfare of all nations. These globalists point out that it is a great chance for the global expansion of the democratic–liberal system and for the development of a global capitalist economy that would broaden the benefits achieved by the richer countries to humanity as a whole.

On the other hand, an opposing view is held by globalization skeptics (which includes neo-Marxists, neo-Keynesians, and even neoliberals), who negatively value the prominence of the economic sphere on globalization. This trend considers economic deregulation and transnationalization to lead to the emergence of an out-of-control economic system, completely unattached to the world of the real economy. As argued by Strange, the economic world is emancipated from the real economy, and speculation becomes the driving force behind *Casino Capitalism*. Essentially, detractors suggest that globalization involves the extension and imposition of a global capitalist economic system and the western sociocultural model, which continues to broaden the difference between rich and poor.

The debate is still open. Richard Haass (2020) introduces a key question about the future of globalization: "The question, though, is whether globalization has peaked–and, if so, whether what follows is to be welcomed or resisted." In recent years, a growing number of governments and people around the world have come to view the phenomenon of globalization as a real risk (terrorism, pandemic, refuges, etc.), so there is an increasing opposition that encourages attitudes toward *deglobalization*. The American withdrawal from the Paris Agreement in 2019 and the World Health Organization in 2020 under the Trump administration are good examples of this new phenomenon.

2.4 New Configuration of the International System

The changes that have taken place between actors, interactions, and processes in recent decades have had a strong effect on the global configuration of the post–Cold War international system. In this sense, these are the two most relevant conclusions that can be reached: (i) the international system is moving towards a scenery of *intersystem transition*, moving on from the old bipolar Cold War system to a new multipolar stage, and (ii) as a consequence of this transition, we can start to notice the first traits that define the *new global configuration* of the international system.

Intersystemic transition

It is possible to consider the old configuration of the international system as a transitional structure, from an old and simplified bipolar model, characteristic of the Cold War, to a new, more polarized, and complex international order. It is what Dallanegra Pedraza (1998, 2010, 2012) calls a process of *intersystem transition*, which is to say, "the period between the 'decadence' of a system and the 'emergence' of another." This systemic reconfiguration and the definition of its main traits is a historic process that can take decades to consolidate. The closest historical example is the transition from the multipolar system (born in 1815 with the Vienna Congress, whose decadence started with World War I (1914–19)) to the bipolar system that emerged in the period between 1945 and 1947 (after World War II and before the Cold War). About 25 years went by during this transition process, which included two world wars and an economic crisis (1929–30) (Dallanegra Pedraza, 2010, 2012). This transition between different stages of history represents the exhaustion of an established order and the emergence of a new one and is characterized by generating structures in which the actors, phenomena, and relations of the emerging order coexist with those of the decaying structure. These are multidimensional processes, affecting the deepest structures of politics, economy, society, and culture.

The magnitude of the change produced by the processes that manifest in the international system since the end of the Cold War and the transition to a new world order are central to the actual process of intersystem transition. The typical bipolar structure that identified the Cold War period gives way to a new configuration that still shows uncertain characteristics. In the '90s, some thinkers foretold the beginning of a new unipolar era, led by the United States, seen as the victorious power of the Cold War contest. However, the events that took place during the first decade of the 21st century meant the apparition of new interpretations of the international configuration, giving way to other opinions that speak of a new historical period characterized by multiple changes that have taken place in the political, economic, social, and cultural spheres by the rise of strong processes and international, supranational, interregional, and transnational dynamics as well as the rising of a new and multipolar world order.

The changes currently operating within the international system are a product of the changes to the main parameters of international relations, leaving behind traits of the old configuration while new ones are emerging. It is therefore the rupture of many of the previous structures on which the international configuration was based and the emergence of new phenomena, processes, and actors that are shaping a new international system.

Towards a new global configuration

This analytic revision of the international system has allowed us to identify new and incipient systemic factors that are starting to describe some of the characteristics that this new international configuration will presumably have. Even though the intersystem transition within our current international system is still an ongoing process and many of the systemic parameters are yet to be defined, it is possible to identify some initial characteristics of this new world order:

(i) The existence of a large variety of State and non-State actors, some traditional and others innovative in international relations, that coexist in a disorderly situation and are highly interdependent among themselves.
(ii) The acceleration and intensification of the interactions between different international actors, consolidating conflicting and asymmetric links, but, at the same time, augmenting cooperative and competitive interrelations.
(iii) The emergence of a new, more numerous, nonhierarchical, and complex global agenda of topics, where new issues that were displaced or subordinated by bipolar logic have re-emerged with great force, such as environmental issues, poverty, social inequality, migrations, world governance, or financial volatility.
(iv) The appearance of new processes in the international system that deeply affects the international reality and its most essential parameters (political, economic, technological, and social).

The new global configuration conditions ISR and, at the same time, is conditioned by all that happens there. Here lies the relevance of knowing what is happening within the ISR and explaining how all this affects the whole international system.

Chapter 3

ACTORS

Actors are considered to be the central units of the international system, which is where they exist and interact. They are, essentially, dynamic parts or elements of the system that constantly interact between themselves through a large variety of ways (conflict, cooperation, competition, etc.) with the goal of satisfying their personal interests. Currently, we can identify a large variety and quantity of actors, both State (countries) and non-State (companies, international organizations), that coexist and interact as parts of the international system.

To be considered as an international actor, one must have the capacity of creating or participating in relations that are internationally significant for the international system as a whole. This means that we must not include in this concept groups or entities that, despite having played an important role in the international scene in a particular historical moment, have lost that prominence as a consequence of the changes that took place in international society, or those that are somewhat relevant in a field but not in others.

In the current historical context of the beginning of the 21st century, we can differentiate a vigorous group of actors with a need to participate in and influence the field of ISR. Some of them have a long tradition in the production and management of scientific knowledge, while others are looking to become more deeply involved in the dynamics and processes of STI for the very first time.

3.1 Old and New Actors

Among the most relevant changes in the international system in recent decades, there are changes in the number, power, function, and behavior of actors. In our current international context, the modern Nation-State has started showing signs of fatigue, while the international prominence of new non-State entities is growing. This same reality can be observed in the subfield of ISR, where traditional actors are trying to adapt to the new relevance of scientific knowledge, while new actors, both State and non-State, are starting to show interest in participating in this increasingly dynamic system. All of them are going through a strong reconfiguration of their old functions, their roles as scientific and social actors, and in the way in which they interact with their peers and with the rest of the actors within ISR.

Currently, the prominence acquired by STI in the international sphere has transformed a task carried out by a group of isolated people, locked away in their workplaces and with limited economic possibilities, into a very strategically valuable activity that is planned and organized by multiple international actors. Now, this activity is carried out by highly skilled research teams, with well-defined purposes and goals, with significant financial resources, and, in some areas of knowledge, with large and complex facilities. Initially, the development of scientific knowledge was based on public policies mainly designed by governmental actors that were then executed through government structures. However, in recent decades, significant changes have taken place that have substantially modified the decision-making process related to STI, which orient the mechanism towards a scenario where more actors play a significant role in the production, intermediation, and governance of knowledge.

This new reality implies that, as the importance of knowledge as a strategic resource grows, so does the number of actors interested in it. For this same reason, if the relationship between scientific knowledge and international actors will determine dominating interests, then identifying who are the intervening actors and the characteristics they present will be very pertinent to understand the type of ISR and international systems we will have in the near future. For this purpose, it is necessary to distinguish which are the main actors that have an active role in the processes of knowledge and to be able to know their objectives and interests as well as their dynamics of interactions with other stakeholders.

3.2 Universities

From its remote and somewhat confusing beginnings in the Far East, through the creation of the first institutions in medieval Europe, to its more recent formalization and structuring, the *university* has always been considered as a central institution in the international system. It is an actor by its own right, linked to a vast array of national and international actors at different levels, as well as a diversified agenda of social, political, economic, educational, technological, and cultural topics.

In the field of ISR, the university has played a very significant role throughout history, not only as the actor naturally responsible for the production, systematization, and transmission of science and technology but also by mobilizing professors, researchers, and students from all around the world across borders. However, and despite its great fame and reputation, in the modern international context, the university faces great pressures and complex challenges that threaten its existence and will test its capacity to adapt to and survive the new environment of the 21st century.

Historical particularities

Throughout history, the university has displayed a series of unique characteristics. These particularities largely explain its historic path and, at the same time, allow us to anticipate its response against the many challenges it must face in the international system.

The first point is considering the university as the *traditional actor* of scientific knowledge. It is an institution with the specific task of creating scientific knowledge and, because of that, what really defines the nature of the university is its historic role as the creator and transmitter of new knowledge. A pioneer in this field such as John Henry Newman points out that the main purpose of the university is knowledge. In the same line, Michel Foucault (1972), the French thinker, considers that the university is a part of society that tries to act on humanity's will to understand the truth in all circumstances of life and the universe in which we are living. Finally, Philip Altbach, Liz Reisberg, and Laura Rumbley (2010) complement this point of view by pointing out that the university is "the primary center of learning and reservoir of accumulated wisdom." In this sense, we understand that the university has historically fulfilled the two main tasks that had been assigned to it: *producing knowledge* and *educating people*.

A second point that has historically characterized the university is its evident *universal vocation* and its open promotion and characterization as an *international agent*. It must be considered that the university has been, all throughout its history, a globalized institution where professors and students shared a great variety of nationalities and where the flow of international students was the norm. That is because, despite the existence of other models of institutional organization, modern universities derive from the western medieval model, especially promoted by the University of Paris, which means that the organizational model of contemporary universities calls back to a common tradition that is an element of internationalization (Altbach, Reisberg, and Rumbley, 2010; de Wit and Altbach, 2020).

At the beginning of the 19th century, with the advent of European nationalism, the university was subjected to strong pressure from the States, who wanted to nationalize its functions and turn it into an instrument for internal homogenization. Thus, the university was forced to provide the citizen with a formal education that would transform specialists into worthy patriots who would defend national interests. Essentially, States and universities reached a silent agreement, by which the latter acquired some autonomy and certain privileges in exchange for creating knowledge and qualified personnel that would contribute to the economic and military development of the nation (King, 2004). This agreement helped to protect the perspective of the university as a universal and international force (as this is the source of knowledge), despite it providing essential services for national agents. The rise of globalization has allowed those national restrictions to begin to be overcome through the growth in the number of foreign students and the creation of a global market for higher education. Altbach, Reisberg, and Rumbley (2010) sum it up by saying, "There is an institution that has always been global and that continues to be a powerful force in the world after half a millennium. With its roots in medieval Europe, the modern university is the center of an international knowledge system that embraces technology, communication, and culture."

The third and last point allows us to understand one of the key points behind the university's thousand-year existence, its *flexibility* and *capacity* to adapt to the innumerable political, social, economic, and cultural changes throughout its thousand-year history. Throughout its history, this institution has shown an enormous capacity to overcome an infinite number of contexts, actors, and phenomena that have passed through it. While other actors emerged, reached their peak, and then simply disappeared, the university remained steady over time, enduring all kinds of changes and maintaining its traditional role as an actor in charge of generating and transmitting scientific knowledge. Its durability can be seen in the fact that, from all the institutions established in the Western world up to the year 1520, only 85 still exist nowadays, from which we point out the Catholic church, the British Parliament, some Swiss cantons, and more than 70 universities (Forest and Altbach, 2006).

The majority of specialized authors agree in pointing out the flexibility and capacity to adapt the university shows against the external context with which it has to interact. According to Barbara Sporn (2006), "universities have become the oldest type of organization in the world, surviving turbulent periods and epochs with different values, beliefs, and cultural norms." This suggests that the administration and governance of higher education institutions have unique characteristics that allow universities to adapt to all kinds of environmental changes. Peter Scott (2002) adds another nuance by pointing out that "what the title of 'university' strongly shows is not its permanence, but rather its institutional adaptability. Universities have shown their capacity to renew themselves, often beyond any recognition."

Ultimately, the centuries-long existence of the university has been characterized by its role as a creator of scientific knowledge, by its universal and international vocation, and by its capacity to adapt to the changing times, which has allowed it to go on occupying a place of privilege as a social and international agent of knowledge.

New roles and challenges

The strengths the university has shown throughout its history are its main assets to face a current international context, where new actors and interests have started to challenge its traditional functions. Roger King (2004) describes the new international context this institution must face very clearly: "the decline of sovereignty and influence of the territorial states, and the growth of international and supranational jurisdiction, alongside the increased globalization of world economy, also heralds, if not the end of the University, then its profound transformation."

The challenges faced by the university in this new international context are multiple and complex: The three most relevant ones seem to be (i) its competition against new actors, (ii) the reduction of its margins of autonomy, and (iii) the assignment of new roles.

New actors

The first of these challenges faced by the university in our current international system is the proliferation of new international actors that aspire to participate in and directly link themselves with scientific knowledge through their own production, transmission, and application. It is what Boaventura de Sousa Santos (2005, 2015, 2018; de Sousa Santos and Meneses, 2019) calls the university's *crisis of hegemony*, meaning it no longer is the only actor in the production of research due to the emergence of new challengers that carry out the same functions that, for centuries, were almost exclusive to the university. In this sense, experts almost unanimously consider that this new international context impinges on the traditional monopoly held by the university in the creation, management, and transmission of scientific knowledge. Immanuel Wallerstein (1996) openly questions if, in the next 50 years, the university will go on being the main organizational base of academic research or if other structures (research institutes, centers for advanced studies, epistemic societies, companies, etc.) will take its place. In this light, many experts are already talking about a new "polycentric" (Innerarity, 2011) or "pluricentric" (Todt,

2021) context, where the university will have to share roles and functions with new actors.

Autonomy

A second challenge the university faces in the new international system is maintaining certain *margins of autonomy and independence* from the advances of other actors. What Boaventura de Sousa Santos called *institutional crisis* is the growing pressure on the university to adapt itself to efficiency and productivity criteria of a business or social responsibility nature. The revaluation of STI as a strategic resource is increasing the attention and the interest from the environment (economic, political, and social) in scientific knowledge and, in particular, is increasing the demands and pressures on the university itself. Breton and Lambert (2003) describes this situation very well when he points out that "universities are now subsumed to the economy and the market, losing the autonomy that they enjoyed at other times, to join knowledge production networks in which the academic decisions begin to be taken from other motivations." Even areas such as the strategic-military or ecological spheres that might seem distant from the university context are starting to pressure the university in search of specific solutions to their own problems. The pressures caused by globalization and economic competition are stimulating new types of linkages between the university and the industry in search of new knowledge and innovations. The majority of current debates, both academic and nonacademic, point out the necessity of the university to coordinate its functioning with other international actors (State, company, international organizations, etc.) with the goal of achieving a greater economic, political, and social development. These new ways of connecting and coordinating new actors with the university are still a serious threat to the autonomy in the university's functioning (Scott, 2020).

New roles

The third challenge faced by the university is the *assignment of new roles and functions*. What seems true is that the university is currently being put under great external and internal pressure to assume new functions and roles. In this sense, we can discern at least three demands: (1) the university should deepen its role as the main actor charged with educating people and training future professionals; (2) it should become a source of scientific and technical solutions to the many global challenges faced by humanity. It is argued that no other actor is better prepared than the university itself, due to its history, nature, and objectives linked to scientific knowledge, to answer the complex international agenda that the 21st century must tackle; (3) lastly, it is also charged with a role linked to the need of sharing the final result of all its tasks with other social actors through specific applications, and very specifically, transferring its research to the production sector as a way of contributing to the economic and social development.

Reconfiguration and adaptation

The new international framework of change in which the university exists, together with the challenges and demands it is facing, has forced the institution to make quick decisions that allow it to successfully adapt to this new global context. For that, the university has had to rethink its interests as an organization and begin a process of reconfiguration drawing from its historical flexibility, adaptability, and international vocation as its main strategies. With this objective in mind, the university has started to develop a series of strategic actions oriented in three directions:

- The university has started to look for *new funding sources* that may allow it to continue playing the role of the main actor in producing and transmitting scientific knowledge. Starting in the '70s, and speeding up in recent decades, many universities have seen their public budgets reduced, which has threatened the institution's very existence and has forced it to look for alternative funding to survive. For that, the university has developed new strategies, including diversification, privatization, and the commercialization of the products and services that higher education generates and can offer to the market (Knight, 2006). It is what Sheila Slaughter and Larry Leslie (1999) have called *Academic Capitalism* and represents the use by the universities of their only real assets, the human capital of its academics and the knowledge generated by them, with the purpose of increasing its income.
- The university has also begun a strategic process of *internal restructuring* in the search for higher levels of efficiency and modernization in at least four areas: (i) in the global administration and management structure, incorporating new ways of management from the field of business; (ii) in the academic offerings, adapting the academic product to the needs of the market; (iii) in research, nudging the development of research to those specific fields where there is a larger interest from the new sources of funding; and lastly, (iv) in its own infrastructure, with a strong commitment to enlarge and modernize its facilities (Jarvis, 2001).

- Lastly, a *reconfiguration of the academic strategies* has also taken place, with the objective of developing a more attractive product in the market of higher education. Among the many actions taken we can highlight (i) strengthening and adapting its academic offerings to the new socioeconomic realities; (ii) raising the international profile and reputation, seeking to attract new talented professors and students; (iii) stimulating the arrival of training and research projects and funding sources; (iv) raising the quality of its education to offer a better academic product; (v) diversifying its students, faculty, and personnel to show intercultural skills and competences; (vi) establishing strategic alliances and networks with international actors; (vii) linking themselves with local stakeholders to establish cooperative links, and (viii) stimulating research and knowledge production on a global level (Knight, 2006).

In essence, it is a reconfiguration of the university's role through strategies that avoid the trend of marginalizing it from the process of production and distribution of scientific knowledge. As Forest and Altbach (2006) point out, the redefinition of the university is an absolutely necessary condition to consider it as one of the main actors of scientific knowledge in the 21st century: "Nowadays, the university's purpose is being questioned and its structure and functions reconsidered. Unless the university responds to the internal and external criticisms and pressures by carrying out thorough reforms, it will be impossible to build universities that can meet the needs of the new era."

3.3 Nation-States

The Nation-State has been, in the last three centuries, the main actor in the field of international studies due to its prominence and crucial role in the world order. However, the new international context has forced Nation-States to adapt to the changes, face the increasing relevance of non-State actors, and develop strategies allowing them to continue playing a leading role in the international system.

In the field of ISR, the interests, performance, and influence of the Nation-State are also threatened by the proliferation of new actors interested in STI. Historically, scientific knowledge has been tightly linked with the Nation-State: from the beginning, where a small group of scientists received funding to pursue a specific national interest, through the boom of Big Science during the post–World War II period, until the current development of complex national plans and public policies on STI. Even though States currently retain decisive weight in shaping policies related to STI, the revaluing of scientific knowledge in recent decades has increased the interest of new stakeholders that are now competing for its production and control.

The emergence of Big Science

At the beginning of the 20th century, the Nation-State showed little interest in the funding of science, which basically consisted of individual, rather modest projects oriented to the achievement of certain national objectives. This budding link between the Nation-State and scientific knowledge radically changed in the '30s with the emergence of a new way of producing and managing science. Before and during World War II, we saw the emergence of what would be called Big Science,[1] which represents the most decisive step taken by a Nation-State to develop new scientific knowledge through strong public investment in science.

The most representative endeavor (though not the first) in this new association between STI and Nation-States was the Manhattan Project,[2] commissioned by the American President Franklin D. Roosevelt, which ended with the development of the first atomic bomb.[3] The Manhattan Project became the first step to the consolidation of the links connecting the State with scientific knowledge, which started to strengthen through the planning of public policies that promoted the development of science and technology.

The pioneering document on scientific and technological policies was the report presented by Vannevar Bush and a group of elite scientists to the president of the United States in 1945, titled *Science, the Endless Frontier*, whose main argument emphasized the funding of basic research as a dynamic principle of the creative process and the transfer of knowledge to the social environment. The report's original objective was to highlight the key role STI was starting to play in the global postwar context; however, it ended up becoming a source of inspiration around the world for the development of public policies linked to science and technologies throughout the second half of the 20th century, in the light of what was called the *linear model*.

---

[1] A name given in 1961 by the American nuclear physicist Alvin Weinberg to those scientific projects that are considered multidisciplinary, multinational, and multiannual.
[2] The project involved an investment of $2 billion at the time, employed 125,000 scientists, and operated more than 10 research centers (Kukso, 2010).
[3] The US president F. D. Roosevelt was assisted by many scientists of note of that time, such as Albert Einstein and Leo Szilárd, and the project was led by prestigious researchers such as Robert Oppenheimer, John von Neumann, and Enrico Fermi.

Thus, a new period started in the link between science and the Nation-State, a result of the renewed interest that science aroused as a generator of wealth, social well-being, and political power. As Kusko (2010) rightly points out, "For the first time, and with an upward and sustained pace, the scientific world joined the political world. Presidents, governors, and senators saw in it a strong force able to shake the game board (and the world) with its discoveries. From then on, they understood, warfare would not be settled in battlefields; it would be defined, rather, in laboratories."

The reason why these States started to more strongly back science was based on the idea that scientific knowledge provides the basis for true progress, both economic and social, even though there was no *a priori* certainty that any immediate changes would take place. This view granted science instrumental value for achieving social well-being, although it had to be complemented with other public policies, such as investment in basic education, human capital, fundamental research, and scientific and technical infrastructure. The underlying idea was to understand science and scientific knowledge as the motors of economic and social development, for which it was necessary to sustain large investigations with large public budgets.

To carry out these strategies, Nation-States designed public policies to reach these objectives. Thus, the concept of *science and technology policy* or *science policy* was born, understood as the set of policies that States adopt with regards to science and technology. Essentially, it should be defined as government intervention in the economy to support scientific discoveries and the development of technological solutions (Chaminade and Lundvall, 2019). In this sense, it is understood as an analogous term to other public policies, such as economic policies, educational policies, or industrial policies, which expresses a scope of public decisions demarcated by a specific object, in this case, science (UNCTAD, 2020).

During the period between the Manhattan Project to the end of the Cold War, Big Science, defined by its enormous investments, enormous infrastructures and bureaucracies, and its multitudinous scientist groups, was the way in which the Nation-State was linked to scientific knowledge.

Post–Cold War era

Despite the criticisms, and even rejection, science and technology garnered during the '60s and '70s, when many groups and social and academic movements raised countless concerns about the beneficial and irreversible character of science, States continued to pay particular attention to the generation of scientific knowledge through large investments in public policies aimed to that effect. Neither did the end of the Cold War diminish the States' interest for scientific knowledge, as it might have been thought *a priori*; on the contrary, it multiplied the attention paid to it, but shifted focus onto other areas aside from the strategic-military domain.

In the face of the strong changes that took place in the post–Cold War international context, Nation-States went a step further in consolidating their interest and link with scientific knowledge, maintaining a commitment sustained by its development through new strategies that collaborate with their *national interests*. These strategic actions have been shaped into the planning of public policies linked to the development of STI, which the majority of States have implemented in the last few years with the objective to increase their investment in the scientific and technological sectors. The most prominent cases include the following:

– The Chinese *Outline of the Medium- and Long-Term Plan for National Science and Technology Development* program *(2006–2020)*.
– The European Union's *EUROPE* 2020. *A strategy for smart, sustainable, and inclusive growth (2014–2020)* and the new strategy *Horizon Europe (2021–2027)*.
– The Japanese *The Science and Technology Fifth Basic Plan (*2016–2021*)*.
– The South Korean *Fourth Basic Plan for Science and Technology (2018–2022)*.
– The Russian *State Program for Development of Science and Technology 2014–2020*.
– The Indian *13th Five-Year Defense Plan (2017–22)*.

In general, all these programs establish priorities and courses of action to be followed by States with the express purpose of improving their standing in STI by focusing on the following points:

(i) A strong investment in R&D, which implies public expenses in scientific personnel, researchers, material resources, and infrastructure.
(ii) An approach decided by skilled human resources, carried out in two ways: (a) with investments in higher education and the training of highly skilled human capital, and (b) with the recruitment of the brightest students and professors from other countries through talent attraction policies and offering job positions and scholarships.
(iii) The search for strategic alliances through regional integration processes to establish productive geopolitical and economic relations.
(iv) The signing of bilateral and multilateral trade agreements, considering education and research as potential areas of profit.
(v) Institutional and individual investment to achieve the training of qualified citizens and well-trained workers.

The development of these policies has become, in recent decades, an important fight between States to achieve the highest level of competitiveness in the field of STI. The increasing expenditures in R&D which, in the beginning, were mainly observed in the most developed countries, have slowly started to spread in a large number of other countries; this is because STI is now seen as a necessary strategy to achieve greater economic and social development and also as a mechanism to face the main *Grand Societal Challenges* (Kuhlmann and Rip, 2018). The expansion of STI plans in many low- and middle-income countries reflect the search for a growth strategy that allows for greater development in the medium and long term (UNESCO, 2015).

Sub-State entities

It is also important to highlight that, inside the wide territory of many Nation-States, there are developments of local and/or regional public policies also aimed at the promotion of STI that exhibit their own peculiarities. This happens with *sub-State entities*, also called *noncentral government actors*[4] (Ugalde Zubiri, 2006). These actors, subordinate to the juridical-administrative power of the State, can be seen as international entities able to link themselves with their peers with which they establish cooperation relations. This is possible because they have certain autonomy to develop their own capacities to manage their own policies in areas such as STI.

Although, indeed, these sub-State entities are not international actors that are completely autonomous from the Nation-States, what is relevant for this type of entities is that they have reached important autonomy margins to be able to develop their own public policies, projects, and plans in the field of STI with the express objective of achieving social and economic development in their own geographical area. Sub-State entities are mainly composed of cities and local autonomous regions, which, in many cases, have assumed the control of competencies very directly linked to planning, producing, and distributing scientific knowledge.

As far as regions go, it is possible to find many strategic plans that represent a clear commitment to the growth of STI as a key element in the social and economic development on a regional level. Examples of this are the Basque Country in Spain (Bases Estratégicas y Económicas del Plan de Ciencia, Tecnología e Innovación - PCTI Euskadi 2030) and Québec in Canada

---

[4] These entities are defined by three elements: "The qualities of an international actor; the nature of governments with certain capacities, powers and governmental competencies; and the political-institutional location of their average administrations other than the central state administration (non-central)" (Ugalde Zubiri, 2006).

(Stratégie québécoise de la recherche et de l'innovation 2017–22). In these cases, they are national regions that manage their own STI policies, which are applied to all the administrative jurisdictions in which they have power.

In recent years, the role played by cities has acquired a special relevance. In the current context of R&D-based competitiveness, cities are becoming more attractive geographic units for the development of STI, either due to their scientific installations and infrastructures or to global talent. Following the successful model of Silicon Valley/San Francisco in the United States, many cities are trying to copy the characteristics that turned Silicon Valley into the paradigm of the high-tech and innovative city. This phenomenon has led to the concept of *innovation hubs*, understood as a community that enables the interactions between actors for the development of activities linked to STI. The leading cities are successful because they enable the exchange of knowledge between institutions and organizations in their environment, allow a higher concentration of talents, and can foster more knowledge-intensive economies as they provide an inviting place to work, invest, and research[5] (The Royal Society, 2011).

3.4 Intergovernmental Organizations

Intergovernmental organizations (IGOs) are also active international agents linked to the field of ISR. From the budding ideas of the then president of the United States, W. Wilson, which gave them their debut as an international actor at the beginning of the 20th century, through their boom and expansion after World War II, to their current status as a moderator in international affairs, IGOs have come a long way as actors in the international system. Usually, an IGO is defined as an organization composed primarily of sovereign states, or of other intergovernmental organizations. IGOs are established by treaty or other agreement that acts as a charter creating the group. Examples include the United Nations, the World Bank, or the European Union (UIA, 2020).

In our current international context, IGOs have acquired a relevant place with regards to the generation and the monitoring of the phenomenon of STI. In some cases, by chasing the interest and benefits of their respective member States and, in others, with a wider approach, by seeking the promotion of a more global and balanced distribution of scientific knowledge. Up to the year 2014, there had been 7,757 identified intergovernmental organizations with an active presence all throughout the international concert (UIA, 2014),

---

[5] The ten main innovation cities in the world are New York, Tokyo, London, Los Angeles, Singapore, Paris, Chicago, Boston, San Francisco - San Jose, and Toronto (Innovation Cities Program, 2019).

and many of them with a long tradition with scientific knowledge being an important topic in their work agendas.

A global review of these international actors allows us to establish a primary and important distinction between *organizations specialized* in *STI* and *regional processes*. In the first case, the distinction is related to the organizations that are created specifically with the objective of addressing the phenomenon of scientific knowledge; meanwhile, in the second case, it is related to the so-called regional processes of integration, in which the key element is geographical proximity. This differentiation is key for the analysis of IGOs because the treatment and management of scientific knowledge is completely different between the two types.

Specialized organizations

There is a variety of institutions that count the treatment of scientific knowledge as one of their main objectives. Among them, we highlight the United Nations Educational, Scientific, and Cultural Organization (UNESCO), which is the specialized organization within the United Nations that carries out studies, research, and consulting work concerning education, science, and innovation.

UNESCO, as an international organization that depends on the United Nations, and especially tackling the sphere of science, education, and culture, has developed intense activities trying to analyze the major changes that have taken place in the field of STI in recent decades and the impact that they have on the international system. This is proven by the numerous reports and research projects it has presented in recent years, where knowledge has always had a prominent position.[6] From its very creation, the statutes of UNESCO clearly state the objectives and functions of this intergovernmental organization and its links with knowledge and science: "Article I: Purposes and functions: (c) Maintain, increase and diffuse knowledge: The purpose of the Organization is to contribute to peace and security by promoting collaboration among the nations through education, science and culture. And this can be done by disseminating knowledge and stimulating intellectual activity" (UNESCO, 2006).

In its more than 60 years of activity, UNESCO's task has been essentially focused towards three areas: (i) mobilizing the international scientific community around topics and challenges considered key for humanity and

---

[6] Although there are many project reports published by UNESCO, the most relevant from the last decades are *Towards Knowledge Societies* (2005); *Science Report 2010* (2010a); *World Social Sciences Report: Knowledge Divides* (2010b); and *Science Report 2015: Towards 2030* (2015).

requiring international cooperation; (ii) consulting in policies and strategies for the development of capacities in the field of STI; these activities include contributions to the development of public policies, educational stimulus packages, and the promotion of participation in the world of science; and (iii) focusing its activity specifically in developing countries and the poorest nations of the world, where it seeks to motivate and stimulate the development of science and technology as part of a more homogeneous global progress.

The *Organization for Economic Co-operation and Development* (OECD) has neither the specific function of UNESCO nor its universal character but in recent decades, it has also developed a strong activity linked to STI as a result of the renewed economic value scientific knowledge has acquired in the international system. As an IGO, it has become a *thinking tank* with regards to science, technology, and development, with the purpose of guiding the creation of scientific policies in its 34 member nations.[7]

Finally, in recent decades, we have seen the resurgence of IGOs that are starting to have a more prominent role in the field of STI, such as the *World Trade Organization* (WTO), the *General Agreement on Trade in Services* (GATS), or the *World Intellectual Property Organization* (WIPO), which have transformed into new and important global forums for discussing the organization and the management of scientific knowledge linked to the economic sector.

Regional processes

Other international organizations that are linked with STI are those known as *regional integration processes*, which link countries in the same regional area to establish mutually beneficial relationships between all its members. Normally, regional processes establish joint public policies in specific areas with the intent of cooperating and obtaining better results than if they acted by themselves. In this sense, the field of STI has always been a privileged space in which countries have found a safe place to cooperate; thus, it is not strange to see how many regional processes there are carrying out important joint developments in the field of STI, giving it a central role in the successful future of regional processes.

The European Union has been one of the regional groups that have shown the most concern and interest in approaching scientific knowledge as a new strategic resource, boosting in recent decades many joint efforts to put STI in

---

[7] Among the main reports of the OECD, we highlight *Science, Technology and Innovation Indicators in a Changing World: Responding to Policy Needs* (2007); *Science, Technology and Industry Scoreboard 2011: Innovation and Growth in Knowledge Economies* (2011); *OECD Science, Technology and Innovation Outlook* 2016 (2016); and *OECD Science, Technology and Industry Scoreboard* 2017*: The Digital Transformation* (2017).

the center of their regional policies.[8] The quinquennial Framework Programs for Research and Technological Development began in 1984 and have continued until current days (now on their ninth iteration, Horizon Europe 2021-2027), with a net investment exceeding €190 billion. The European Commission (2010) is committed to *smart* economic growth for the European Union, representing "the consolidation of knowledge and innovation as a promoter of future growth."

In Latin America and the Caribbean, the efforts of distinct regional processes are also significant: The Southern Common Market (Mercosur), has prioritized cooperation in STI from the beginning, with the creation, in 1992, of the Reunión Especializada de Ciencia y Tecnología (RECYT), and later, in 2005, with the creation of the Reunión de Ministros y Altas Autoridades de la Ciencia, Tecnología e Innovación (RMACTIM) as a policy-making body, which aims to provide, strengthen, and expand opportunities for scientific and technological collaboration between Member States. For its part, the Union of South American Nations (UNASUR) approved a year after its creation, the *South American Council of Science, Technology, and Innovation* to stimulate the scientific cooperation between its members. In addition, the members of the Caribbean Community (CARICOM) established in 2014 their first regional program for scientific cooperation, called The Strategic Plan for the Caribbean Community (2015–2019), where they seek to agree on regional STI policies that avoid duplication and promote synergy in regional scientific research.

In other geographical regions, we can see similar processes: in Asia, the Association of Southeast Asian Nations (ASEAN) has launched a Plan of Action on Science, Technology, and Innovation (2016–2020), where sustainable development and social inclusion are a priority and the focus is on areas such as drinkable water, renewable energy, social innovation, and the development of green technology; in Africa, the Economic Community of West African States (ECOWAS) has made progress on cooperative agreements in STI, firstly through Africa's Science and Technology Consolidated Plan of Action (2005–2014), which aims to establish a regional network of excellency centers and to stimulate the mobility of researchers throughout the continent, and last through the plan of action titled ECOWAS's Vision 2020, with the main objective being to establish a roadmap to improve governance and STI cooperation in the continent.

---

[8] One of the founding pillars of the current European Union was EuroAtom, created in 1957 by the Treaty of Rome, which became the European public body charged with coordinating the nuclear power research programs.

3.5 Nongovernmental Organizations

Nongovernmental Organizations (NGOs)[9] have also become important actors in ISR, promoting STI as a tool to solve humanity's most relevant problems, such as humanitarian crises, environmental problems, or poverty. Although the historic precedents of NGOs go back to the 19th century, they were finally formalized as an international actor in the United Nations Charter in 1945, where they were identified as organizations independent of the influence of governments and with different objectives to the companies. An NGO is a legally constituted organization created by private persons or organizations without participation or representation of any government. The term originated from the United Nations and is usually used to refer to organizations that are not conventional for-profit business. NGOs can be organized on a local, national, or international level (UIA, 2020).

Rise of the NGOs

These new actors, which mostly concentrate their activity on the completion of a mission or goal, have grown extraordinarily during the latter half of the 20th century, becoming a major player in the current international system. The evolution of NGOs has been extraordinary: from a scarce 1000 NGOs at the beginning of the 20th century to the 10 million registered today. Just in India, we can find 3.3 million NGOs, and in the United States, there are another 1.5 million (Nonprofit Tech for Good, 2018). This numerical growth was accompanied by an even larger economic development: in 2014, NGOs received more than a billion dollars in donations, and it is expected that this number will increase to more than 2 billion in 2020. At the same time, the number of employees, benefits, and countries of operation have not stopped growing. Among the outstanding examples, we find *BRAC* (100,000 employees, 60 billion in income, and 15 countries), *Doctors Without Borders* (36,000 employees, 1.6 billion in income, and 70 countries), and *Oxfam International* (4,500 employees, 83.2 million in income, and 90 countries). This spectacular growth in their role as international actors linked to STI has increased the influence of NGOs when participating in the development of public policies at a local, regional, and global level.

As also happens with other international actors, this relevance has also been translated into the apparition of international rankings that classify

---

[9] In some countries, the term NGO is applied to organizations that in other places would be called NPO (Non-profit Organization) and vice versa.

NGOs according to their main characteristics and performance. Since 2012, NGO Advisor has published a ranking that, in 2018, has reached its sixth edition and where the 500 most important NGOs of the world are identified (Global_Geneva, 2018).[10] Generally speaking, NGOs vary greatly among themselves in size, geopolitical location, sector, objectives, political or philosophical standing, interests, source of funding, and how they operate. The most common way of classifying NGOs is by their *scope of application* (including human rights, the environment, health, or development) and by their *field of operation* (i.e., local, national, regional, or international). Even though NGOs deal with a great quantity of economic and social topics, in recent decades, the largest NGOs have focused especially on topics linked to development, the environment, and human rights. Among the most recognized and outstanding NGOs, we have *Amnesty International* (Human rights), *International Transparency* (Corruption), *Greenpeace* (Environment), *BRAC and Oxfam* (Poverty), *DRC and Mercy Corps* (Refugees), and *Doctors Without Borders and Care International* (Humanitarian Aid).

Promotion and uses of science, technology, and innovation

NGOs' open interest for scientific knowledge manifests itself in at least two ways: (i) the development of STI and education are themselves an objective of NGOs, who understand the impulse and promotion of scientific knowledge as a key element of scientific development; and (ii) NGOs have stimulated the use of STI as the most appropriate tool to tackle the most relevant social and economic topics of the current global agenda.

Impulse and promotion of science, technology, and innovation by NGOs

The promotion and dissemination of topics related to STI are one of the main tasks of NGOs. The historical precedents can be traced back to the 17th century with the appearance of the *Royal Society*, which was a pioneer in the promotion of science for the benefit of humanity. Although the majority of NGOs dedicate their time and effort to social topics such as poverty, environment, or humanitarian crises, the number of NGOs working in the specific field of STI is increasingly growing up (there are 7,191 such NGOs in

---

[10] The top 10 of the 2018 top 500 NGOs ranking are as follows: (1) BRAC, (2) Wikimedia foundation, (3) Acumen Fund, (4) Danish Refugee Council, (5) Partners in Health, (6) Ceres, (7) CARE international, (8) Médecins Sans Frontières, (9) Cure Violence, and (10) Mercy Corps (Global_Geneva, 2018).

the world). Many NGOs seek to link themselves to UNESCO to build synergy for carrying out joint activities, and, at the beginning of 2018, there were 389 NGOs in the world with some sort of formal or informal agreement with UNESCO for the promotion of such topics. Currently, it is estimated that, in the United States, 1% of all NGOs are focused on topics linked to scientific and technological development.

Use of STI as a tool

NGOs have become one of the actors that have encouraged the use of STI to tackle the complex international agenda. As working with topics linked with poverty, taking care of the environment, the lack of resources of humanitarian aid is part of the mission and objectives of NGOs, scientific knowledge has become a primordial tool for the work they carry out to improve the infrastructure, communications, and logistics of their fieldwork. At the same time, applying scientific solutions to specific issues of the populations they are helping in the fields of health (new vaccines or treatments), food insecurity (new sowing and irrigation techniques), or natural disasters (using satellite systems for their prevention) (Makri, 2013).

Main NGOs in science, technology, and innovation

The work carried out by NGOs in the field of STI is expanding at the same time that scientific knowledge is becoming increasingly relevant in the international system. In this sense, it is important to highlight the work some NGOs are doing around the world:

- *The International Council for Science:* Founded in 1931 as the *International Council of Scientific Unions*, it took its current name in 1988 and its members are national scientific bodies (122 bodies representing 142 countries) and international scientific unions (30 members). Its mission has always remained unchanged as the search for international cooperation for the advancement of science. In recent years it introduced its Action Plan (2019–21), where it established a long-term vision for the organization, where we can identify three priority axes: (i) stimulating international cooperation in research, (ii) increasing the usage of science for the design of public policies, and (iii) supporting universal access to science.
- *The Royal Society:* It is considered one of the oldest and most prestigious scientific associations in the world. It was founded in the 1660s, and its objectives have always been linked with promoting science. Its current

Strategic Plan 2017–2022 has as its main objectives promoting scientific excellence, supporting international cooperation, and encouraging science for everyone.
- *American Association for the Advancement of Science (AAAS)*: Founded in 1840, the AAAS is the world's largest scientific society, with 120,000 members and active participation both within the United States and internationally as a defender of scientific development. Its main objectives are the promotion of science and scientific collaboration, the defense of scientific freedom, stimulating education, and the dissemination of science.
- *International Organization for Science and Technology Education (IOSTE)*: Founded in 1979, the IOSTE's objective is to promote dialogue, discussion, and reflection on the importance of science and technology education. This NGO's main mission is based on the idea that science and technology must be a vital part of the general education of society. For its broadcasting, the IOSTE develops periodical regional symposia, where various actors (governmental sectors, academia, and private) discuss education and dissemination of STI.

3.6 Transnational Companies

The incredible development and transformation of the economic sector in recent decades, triggered by the scientific and technological revolution and by changes in the capitalist system, have strengthened the position of *transnational companies* as central actors in ISR. Although large multinational companies have had a prominent role throughout the history of international relations, in the last part of the 20th century and the beginning of the 21st century, they have become critical actors within the international system. They have grown exponentially thanks to the high independence they have from State, intergovernmental, and/or supranational tutelage and to the lack of systemic limits that control their operation, allowing them to increase their maneuvering room and reinforce their power on the global stage.

STI has always been an important element for companies; however, in recent decades, the link between the private sector and knowledge has strengthened and has become a necessary factor for the operation and success of any transnational company. In the new international context, scientific knowledge has transformed into a resource with high commercial value, and for this reason, companies have started to deploy many strategies for its development and appropriation.

Economic interest

In the modern economy, the change in the development model is transforming knowledge into technology and innovation, which is now the only competitive advantage for companies (Porter, 1998, 2011). For that same reason, in the current capitalist economic system, the interest of companies has shifted to the production of knowledge and innovation, which are now the sole producers of economic benefit.

The main reason behind the company's interest in STI is focused on transforming scientific knowledge into an element of innovation, competitiveness, and economic benefits and on wealth generation. In the current capitalist system, scientific and technological progress, through investment in R&D activities, is a key factor in understanding the generation and accumulation of technological knowledge and resources, positive effects in productivity, and economic growth in the middle- and long-term basis for companies. The change in economic growth is caused by considering these variables as essential motors of economic gain, hence the interest and need of companies to get involved with knowledge (from its creation to its final application) as a strategic element in developing their competitive advantages.

Transnational companies are facing a new international system where scientific knowledge has transformed into the most important factor of production and where the economy is focused on knowledge as the basis of production, productivity, competitiveness, and commercial success. In the context of a *knowledge economy*, this is applied to not only cutting-edge companies but any and all that want to survive in the 21st century's capitalist economic system.

Action strategy

The new *knowledge economy* has forced transnational companies to develop strategies and courses of action that allow them to link themselves with scientific knowledge more closely, safely, and sustainably in the long term. These strategies, whose main objective is assuring companies' access to new scientific knowledge, are based on five main tenets:

- *Talent*: Getting an abundant and well-qualified labor force in the labor market for the main roles in an organization that demands increasingly higher levels of learning. For that, companies demand that higher education institutions develop their academic offerings for careers that are

oriented towards and have a quick outcome in the business and production sector.
- *Research*: Funding scientific research conditioned by the topics and goals they pursue. In many cases, companies have their own research groups but also call on universities to secure new scientific knowledge that can be translated into commercial innovations.
- *Transfer*: Establishing technology transfer policies that articulate the State's investment and universities' operational capacity with the objective of offering results transferable to the companies' productive systems.
- *Patents*: Conserving intellectual property rights to the scientific knowledge gained (it could be a product, a service, or a process), which allows them to derive economic benefits from the legitimate temporary monopoly.
- *Institutionalism*: Creating their own educational and formative organizations, known as *corporate universities*. In this case, companies are assuming the universities' educating role. Although the majority of corporate universities are focused on basic postsecondary education, without academic accreditation and targeted to their own employees and intermediate directive bodies,[11] many of them are also working on opening up to the free market, which means transferring the training methods and certifications of this type of institutions to the general demand of higher education in open competition against traditional universities.[12]

The new relevance acquired by STI in the new international economic system has made transnational companies bet decisively for knowledge through the implementation of a double strategy: on the one hand, they try to control the scientific knowledge generated by other actors with tools such as patents; on the other hand, they also manifest their interest by assuming themselves the role of generators of knowledge thanks to the development of corporate universities.

---

[11] Corporate universities doubled in number from 1,000 to 2,000 between 1997 and 2007 and doubled again between 2007 and 2011 to reach 4,000 universities in the United States and more than twice that number worldwide (Boston Consulting Group, 2013).

[12] The list of the current main corporate universities includes companies in the production sector (General Electric, General Motors, Jaguar Land Rover, Shell), consumer goods (Coca-Cola, Marlboro, McDonald's), commercial (Walmart, Eddie Bauer, Best Buy, Home Depot, Target Stores), financial (American Express), entertainment (Disney, Universal), telecommunications and computers (Apple, AT&T, Microsoft, Xerox, Motorola, Sun Microsystems, Oracle).

3.7 Think Tanks

In recent decades, *think tanks* have reached a prominent position in the international scene as a new actor in scientific knowledge. The unique characteristic of these organizations is that they are mainly focused on scientific research and the production of knowledge related to public policies, seeing themselves as a bridge between knowledge and the power of modern democracies.

What has turned think tanks into a relevant actor for ISR is that their main focus and goal are directly linked with the production and generation of scientific knowledge topics that are then transmitted as technical, reliable, and accurate information for the public and private sectors to be used as inputs for decision-making processes. Mc Gann defines think tanks as follows:

> [They are] public-policy research analysis and engagement organizations that generate policy-oriented research, analysis, and advice on domestic and international issues, thereby enabling policymakers and the public to make informed decisions about public policy. These institutions often act as a bridge between the academic and policymaking communities and between states and civil society, serving in the public interest as an independent voice that translates applied and basic research into a language that is understandable, reliable, and accessible for policymakers and the public. (Mc Gann, 2019)

In this sense, think tanks have become new mechanisms or instruments for the generation and transference of information and agents of political and social change through the creation and accumulation of knowledge and collaboration with diverse international actors.

Predecessors

Despite the novelty of their role as an actor linked to scientific production, think tanks have several predecessors. Following Tello Beneitez's (2013) analysis, it is possible to differentiate three stages:

(i) The first period, from the beginning of the 20th century, where think tanks were considered "true institutions of politic research and established as 'studentless universities' that wanted to educate an elite" (Tello Beneitez, 2013). Their role here still had very blurry limits with regards to other actors such as universities or interest groups.
(ii) A second generation of think tanks appeared after World War II, whose main trait was their closeness to the governments. Their central task was

carrying out international political analysis in the context of the Cold War and assist governments in their decision-making process. RAND (Research and Development Corporation), a society created by the American general Henry H. Arnold in 1948, is considered the first think tank according to our current definition of the term. It was founded at the end of World War II after seeing the importance of technology and research in achieving success on the battlefield and the wide range of scientists and academics, outside the military, who made this development possible.

(iii) Since the mid-'90s, think tanks have developed a huge numerical, on-site and operational force, which has transformed them into one of the world's fastest growing and important international actors: in 2015, 6,846 think tanks were operating in 169 countries around the world. The proliferation, global expansion, and generation of work networks between them have shown the potential these actors have. As pointed out by Mc Gann (2019), "Think tanks have increased in number, but also the scope and impact of their work have expanded dramatically as well." This growth is due, in large part, to the organizational flexibility these actors have, to their capacity to operate in a great variety of political systems, to link up with a broad range of public activities, and to form relationships with a large group of institutions and actors.

Distribution and geographical expansion

Even though the majority of think tanks share the common objective of producing high-quality research and knowledge, their organizational structures, their mode of operation, their audience and markets, and their financial aid varies between institutions and between countries.

Although nowadays it is thought that these new actors are operating in more than 160 countries spread all over the globe, one of their more evident limitations is their unequal geographical distribution and its concentration in western countries. Even though some progress has been made in some regions (Latin America 11%, Africa 15%), there is still a clear geographic imbalance by which North America and Europe lead the scene with 54% of the total. This geographical inequality is amplified if we also consider that, among all the think tanks currently active, almost 30% are located exclusively in the United States.

In any case, in the last few years, and thanks to the advantages offered by the globalized context and technological advances, think tanks have started to execute global expansion and international cooperative actions. In this sense, there have been two main strategies: (i) many think tanks have established cooperative alliances and global networks with think tanks in other countries, which has become an effective mechanism for transferring information and knowledge of international experiences that then can be used on a local level (Mc Gann, 2019); and (ii) some think tanks have positioned themselves in strategic geographic points, establishing centers of operations in different countries outside of their head office. These institutions are organizing think tank networks to collaborate with the development and evaluation of public programs and policies and, at the same time, to act as a nexus with civil society groups at the national, regional, and global levels.

Margins of autonomy

The margins of autonomy that funding allows has become a critical factor for think tanks to be able to become actors with true influence and capacity to act in the international system. The main challenge faced by all think tanks is achieving and maintaining their economic independence, to be able to have the intellectual freedom to pursue research topics as sensible as those linked to the *powers*, and to build knowledge and experience that can be used scientifically as inputs and insights for the political decision-making process.

Even if some think tanks have been able to consolidate strong sources of funding that allow them to work with some autonomy, unfortunately not all of them have independent financial, intellectual, and legal support to be able to work with freedom. Many of them have become simple lobbies moved by personal and/or sectoral interests and that only seek to impose their own agendas, without any interest in generating scientific thought or promoting discussion and reflection about current topics.

The role think tanks can play depends on their capacity to secure higher levels of funding and autonomy: from important actors that will ensure a plural, open, and transparent process of analysis, debate, decision making, and evaluation of public policies to becoming simple euphemisms to refer to groups with particular interests with intentions of imposing their own political agendas on the public debate.

3.8 Epistemic Communities

Another new and relevant international actor within ISR is also one of the lesser-known ones. *Epistemic communities* are networks of experts and specialists that work on the main global issues with the objective of producing knowledge and scientific solutions. In this sense, it shares with universities and thinks tanks a task and a goal which, at the same time, differentiates them from States and companies, who have interests beyond the mere production of scientific knowledge.

In 1992, Peter Haas suggested, for the first time, the usage of the concept of *epistemic communities* to analyze the influence that networks of recognized experts, competent in a specific scientific field, have on the implementation of international public policies. Haas thinks that epistemic communities have a new leading role in the international system as the vehicle through which new scientific ideas are circulated.

Generally speaking, it can be said that epistemic communities share at least four common identifying traits: (i) they have certain beliefs and principles that act as a basis for their actions; (ii) they establish professional and expert judgments; (iii) they have notions of validity; and, (iv) they build a common political agenda (Haas, 1992). Due to their specialized knowledge, epistemic communities have sufficient legitimacy in a specific field as to become a sort of driving force, a promoter, and a conductor of proposals (Caballero, 2009). This definition of epistemic communities does not exclude power relations or existing political interests, and for that reason, some authors prefer to instead use concepts such as *organic intellectuals*, *interest groups*, or *pressure groups*.

New global context

Historically, scientists and intellectuals have shown universal and international vocation and have been predisposed to move all around the planet in search of new evidence, information, examples, or arguments for their projects, and to establish cooperative relations and links with colleagues in the most distant places. However, in our current international context, these interactions have intensified as a consequence of a set of factors that have combined to transform epistemic communities into a powerful actor in the field of ISR. Among these factors are (i) more numerous, more specialized, and more complex topics of the agenda; (ii) the operative opportunities offered by new transportation and communication technologies that have promoted new links; (iii) the relevance that the result of their work has acquired; and, especially, (iv) the weight their expert opinions have acquired.

In the frame of a new global system where it is increasingly important to understand the deep causes of international phenomena, as well as the possible scientific and technical solutions to the great problems affecting humanity, the technical opinion of epistemic communities has become decisive. With this in mind, it is not strange that experts linked to different organizations (IGOs, NGOs, public and private research institutes, corporate study centers, and universities all around the globe) are increasingly required to address global issues and to offer concrete solutions. Oftentimes these experts are part of transnational networks and take part in international conferences and negotiations, where they discuss the state of knowledge, delimit topics for discussion, and propose specific technical solutions. For instance, since 1988, hundreds of scientists all around the world have participated in the Intergovernmental Panel on Climate Change, whose policy reports have given key input in negotiations over global climate change.

Revaluation of expert knowledge

This new global framework allows us to understand the relevance acquired by epistemic communities as an actor of scientific knowledge. Dagnino and Thomas (1999) comment on this influence by pointing out that "the scientific community of developed countries plays an important role in the creation of scientific policies, as well as the social actor responsible for the implementation of the resulting activities." This affirmation raises epistemic communities to the category of a new actor of STI that obtains its influence due to a context that is increasingly characterized by its technical aspects, its complexity, and its uncertainty.

This global repositioning has allowed epistemic communities to transform into international actors with great influence over the process of creation and adoption of public policies. For Caballero (2009), epistemic communities constitute a "driving force, and promoter and leader of proposals that acts as a sort of power throughout the whole of an evolutionary process." According to Haas (1992), epistemic communities not only actively participate in the processes of scientific knowledge creation but are increasingly able to influence the decision makers. This influence is not limited to the mere role of consulting, but it extends to the completion of coordination tasks in the process of creating policies linked to STI. Following the analysis carried out by Adler and Haas (2009), it is possible to identify four main phases where epistemic communities now have a prominent role:

(1) *Innovation of policies*: where epistemic communities exert an influence on the topics of the political agenda through the limitation of the range of a subject's political controversy, in the definition of the main interests, and the establishment of standards and criteria.
(2) *Dissemination of policies*: where epistemic communities use processes of international communication and socialization to promote new ideas and innovations. The members of these epistemic communities not only actively get involved in efforts on a national level but also broadcast their advice on specific policies at the transnational level

through communication with colleagues from other scientific bodies and international organizations.
(3) *Selection of policies*: where epistemic communities have their own voice in the process and must share their opinion with the other actors taking part in the decision-making process.
(4) *Persistence of policies*: where new ideas, once institutionalized, must be monitored and evaluated, which frequently happens thanks to the persistent efforts of epistemic communities.

The analysis of the process of building public policies allows us to discover how epistemic communities have acquired a notable position inside the new context of ISR. This new role is rightly described by Haas (1992) when pointing out that "we can see international policies as the process by which the innovations of epistemic communities are spread in the national, transnational and international levels to become the foundations for new or transformed international practices and institutions and the emerging attributes of a new world order."

3.9 Scientific Diasporas

Finally, there is another actor that can be identified in the field of ISR. *Scientific diasporas* are formed by the transnational population that emigrated to another country but still maintains links with its home country. These diasporas have grown in influence and power in recent decades thanks to the new technologies in transport and communications, which have allowed them to establish themselves as key actors in the generation of positive links between countries in the international system.

The key element in the development of scientific diasporas as an actor is found in the will of a group of migrant scientific workers to promote a process of linking with other scientists of their same nationality in their host country to then establish a contact with the scientific community of their country of origin, with the objective of sharing their knowledge for local development. Scientific diasporas occur in highly skilled personnel networks that dynamically maintain and encourage academic, scientific, and entrepreneurial links with their countries of origin, mainly through new information and communication technologies, thus promoting a flow of knowledge, competencies, and resources. The key here is the implementation of strategies guiding the flow of knowledge, competencies, technology, and other resources between migrant scientists in other countries to generate social and economic transformations in their countries of origin (Tejada, 2012).

As it happens with other actors linked to STI, we can find some predecessors of scientific diasporas collaborating with their countries of origin even before the 20th century; however, these processes were carried out in a more fragmentary way. The changes that have taken place in the international system (mainly the intensive use of information and communication technologies) have created the conditions for scientists and technicians to start producing knowledge on a global scale in the frame of physical and virtual cooperation networks. In this way, teams of researchers, academic and industrial, public and private, operate in systems of dynamic relations based on the exchange of knowledge between the country of origin and the destination of the diaspora, creating a linkage matrix called the *brain network*.

From brain drain to brain gain

Scientific diasporas are seen as a way to benefit from the presence of national scientists abroad, as the talent-exporting countries now have a chance to actively recover the emigrated skills based on the idea that every expatriate can bring virtually something to their home countries from anywhere in the world. In some way, this phenomenon implies a new idea of the relation between qualified immigrants and their places of origin, considering unnecessary the physical return of those who, having integrated themselves in the new knowledge-generating and technology-creating cultures, would be willing to cooperate with the scientific and technological communities of their countries of origin (Meyer, Kaplan, and Charum, 2001).

Scientific diasporas are becoming central actors for the generation of cooperation in STI among countries of origin and the host countries, as well as the generation of effective mechanisms for turning *brain drain* into *brain gain*.

Main initiatives

Currently, many countries are developing public policies aiming to stimulate their scientific diasporas to spread all around the world to connect with domestic actors to take advantage of the knowledge and the experience gained by their fellow countrymen in their stay abroad. China and India are among the countries that are encouraging this type of policy due to the large collection of national scientists and researchers living outside their borders (mainly in the United States and the United Kingdom). There are also representative examples in Africa and Latin America; for example, we can highlight the South African Network of Skills Abroad (SANSA) for its significant contributions to the objectives of South African development; in Latin America, the Colombian Red Caldas, which has been the region's most important benchmark case.

At the same time, international organizations are supporting the development of these networks that link scientific diasporas with their countries of origin. At present, we highlight programs such as the UN's Transfer of Knowledge through Expatriate Nationals, which is mainly focused on creating databases of skilled scientists in foreign countries that would be willing to participate in specific development projects. Another program is Migration for Development in Africa, led by the International Migration Organization (IMO), with the goal of mobilizing the skills of expatriates from African countries for the benefit of the continent (Faist, 2005). The creation of virtual research networks in STI has had a strong impact on the international development agenda, and for that reason, UNESCO, the UNPD, the OECD, the World Bank, and the European Union are promoting the development of virtual research networks for the linking of expatriate scientists and researchers.

Chapter 4
RELATIONS

The new international context characterized by an extraordinary scientific and technological revolution, the expansion of the phenomenon of globalization, and the unprecedented development of information, communication, and transportation technologies is stimulating a constant increase in the relations and interactions between all the actors of the international system, which suggests a future scenario where the number and the diversity of the interconnections will multiply exponentially.

This same reality can be observed in the field of ISR, where the reevaluation of scientific knowledge as a strategic resource and the renewed interest of old and new international actors in STI are shaping a new global structure with new means and a larger quantity of interactions and relations between different actors.[1] Currently, if possible, observing an expansion in the number, nature, and reach of the interactions between the different actors and places of knowledge production leads not only to more knowledge being produced but also to the acceleration and intensification of connectivity.

The main objective of this chapter is to identify the main interactions and relations that are generated between the major actors of ISR and to describe their main characteristics, because, as Strange (1994) points out, a large part of the researcher's job is "trying to unravel the complex network of overlapping, symbiotic or conflictive authorities that can be found in any sector or any time we are asked who gets what."

4.1 Interaction and Relation Mechanisms

The huge contact, link, and interaction network that exists in the field of ISR among the many international actors comprise a huge *network of relations*

---

[1] We must remark on a very subtle but important differentiation between the very similar concepts of *interaction* and *relation*. Interactions are understood as happening on the short term and during short periods of times and allow us to understand the international stage. These actions are called relations when they happen on the medium or long term (Calduch Cervera, 2017).

(Dagnino and Thomas, 1999), which we see represented the main economic and political interests of the actors involved in scientific and technological activities, who can either be producers, consumers, or funding agencies. Within this network, a process of reciprocal influences between all these actors takes place, where these interactions generate a *breeding ground* in which values are shared at the same time that research priorities are established. In the configuration of this relationship matrix, policies are defined and the allocation of resources that define the fields of relevance, the research trends, the specific weight of the different areas of research, the priorities, and even the criteria of quality are decided.

The interactions and relations between different actors in ISR have been profoundly modified in the last few years due to the confluence of a myriad of factors endogenous to the international system (the reevaluation of scientific knowledge, the emergence of new actors, the acceleration and intensification of contacts, and the consolidation of the phenomenon of globalization). The majority of contemporary debates include the idea of an increase in linkages between all actors interested in knowledge and innovation. The most widespread models in academic literature, such as theories on *Innovation Systems* (Lundvall, 1985, 2004; Chaminade and Lundvall, 2019); the *Mode 2* of knowledge generation (Gibbons et al., 1994; Nowotny, Scott, and Gibbons, 2003, 2006), *Mode 3* (Carayannis and Campbell, 2006, 2009; Carayannis, Campbell, and Rehman 2016), the *Triple Helix* (Etzkowitz and Leydesdorff, 2000; Leydesdorff, 2012); the *Quadruple and Quintuple Helix* (Carayannis and Campbell, 2009, 2016), and even alternative points of view such as the *Social Development* (Lander, 2008, Núñez Jover, 2006, 2020; de Sousa Santos 2005, 2015, 2018; de Sousa Santos and Meneses, 2019) all converge on the same point of view: the existence of a larger quantity and intensity in the linkages between all actors interested in STI.

Although it is possible to find diverse ways of classifying and/or distinguishing the relations between international actors in academic literature[2], our goal is to get to know and understand the interactions carried out by all international actors that establish any degree of relationship but that by their nature are specifically linked to the area of STI. Considering the most relevant interactions in international relations, we can divide the way actors

---

[2] Calduch Cervera establishes a classification or typology of interactions according to different criteria: (i) by the number of actors intervening (bilateral, multilateral or global); (ii) by the degree of linkage established between actors (direct or indirect); or, (iii) by the nature of these institutions (political, economic, legislative, cultural, scientific, technological, etc.) (Calduch Cervera, 2017).

in the field of ISR interact into four main categories: conflict, cooperation, competition, and asymmetry.

4.2 Conflict Interactions

In the international system, power relations are natural behaviors and, in particular, in ISR, conflictive interactions between actors are very common. For many experts, the explanation for the existence of conflictive links in the field of scientific knowledge is based on the same antagonistic nature of the relation between science and international actors, due to the dissimilar interests pursued by each one: scientists, under the cover of universal values, try to develop technical and scientific progress of global scope, while countries, companies, and universities defend and fundamentally promote particular interests.

In recent decades, conflictive relations in the field of ISR have grown in number and intensity. Some experts understand it as the logical consequence of the increase in strategic relevance that scientific knowledge has acquired in the new global system, which rationalizes the fight for such a transcendental element in a global context where the fight for resources and particular relations has always been the norm (Quintanilla, 2007; Innerarity, 2011, 2013). However, for other critical specialists, the new context of economic globalization is the cause of the new sources of confrontations, as STI has become the main means of production of the 21st century and the fight inside of the capitalist economic system has focused on the processes of production, appropriation, regulation, and application of scientific knowledge (Núñez Jover, 2006, 2020; Lander, 2008; de Sousa Santos, 2005, 2015, 2018; de Sousa Santos and Meneses, 2019).

In the current context of ISR, the main sources of conflict are settling on two main fields: (i) conflicts derived from the intent of the main actors of the international system to attract and capture the largest possible quantity of world talent, seen as those mainly responsible for the production of STI; and, (ii) conflicts linked with the appropriation of knowledge itself, which is expressed through property rights and patents, through which many actors seek to protect and monopolize the creation of scientific knowledge with commercial purposes.

Fighting for global talent

The first group of conflicts to analyze is that linked to the fight carried out on a global scale to procure the services of *global talent* or *highly skilled personnel*,[3] which

---

[3] Those human resources in science and technology that successfully completed their tertiary education, that is to say, that have at least an university degree or a schooling

are the people responsible for the creation, development, and manipulation of scientific knowledge.

Talent mobility throughout the planet is not unprecedented in history; on the contrary, it is a habitual practice by which scientists and experts have gone wherever the context and opportunities were most favorable for their work. Currently, the international mobility of highly skilled personnel is framed in a global scenery of internationalization for STI and higher education activities, where not only individuals decide to mobilize but also the State, through its public policies, and other international actors such as universities, research centers, and companies, are encouraging those movements through attraction strategies.

This phenomenon is based on the widely generalized consensus that, from the economic point of view, the effects of immigration are beneficial for the host countries. In the United States, more than a third of documented immigrants are considered members of the skilled workforce, and in Europe, there are similar trends, which reflect the needs of those economies. Governments open to immigration help national companies, which become more agile, adaptive, and profitable in the fight for talent; for their part, governments receive income and citizens benefit from the dynamism brought by skilled immigrants. Some research on the net tax impact of immigration in the host countries shows that immigrants contribute significantly more to taxes than the benefits and services they receive. According to the World Bank, increasing immigration in a margin equal to or 3% larger than the labor force in developed countries would generate earnings of 356 billion dollars. In the same line, some economists predict that, if borders were completely opened and workers were allowed to go wherever they wanted, it would generate earnings of up to 39 trillion dollars for the world economy in only 25 years (Goldin, 2016).

On a global level, the key role played by qualified human resources in socioeconomic development and the need to attract them is recognized due to the fact that "in the future, talent, more than capital, will represent the critical factor of production" (Schwab, 2016). For that, the key to this conflict is found in international actors' ever-increasing interest in attracting highly skilled personnel independently of their location, thus stimulating their flow across borders.

---

of at least 13 years in a field of science (knowledge) or technology (applied knowledge), and/or those who do not necessarily have that qualification but are employed in an area of science and technology in a position where tertiary studies are usually required (OECD, ANSKILL Database, 2012).

Historic evolution

As pointed out by Stevan Dedijer (1968), "scientific migration is as old as science," due to the long tradition of scientists and researchers moving to where scientific knowledge and its advances could be found. For a large part of history, this flow of professionals, technicians, and scientists has been called *mobility*, *exodus*, *nomadism*, or *brain drain*, thus becoming one of the main sources of conflict between international actors. The counterpart to the attraction and recruitment of highly skilled personnel is the concern of the loss of those same human resources in their places of origin.

Since the '90s, the international mobility of the most qualified workers has increased and intensified due to the rising global demand for specialists and the extraordinary development of information, communication, and transportation technologies. In the year 2000, the number of migrants older than 25 years with higher education and living in the OECD countries reached 20 million, as opposed to 12 million in 1990. Between 2000 and 2010, the number of immigrants with higher education in the OECD countries increased by 70% and reached a record-breaking 28 million in 2011, which represents an increase of 130% between 1990 and 2010. Four OECD countries (the United States, the United Kingdom, Canada, and Australia) have received approximately 70% of those migrants, while the United States had got close to half of the total migrants of the OECD and a third of the global total (OECD, 2013).

Lack of global talent

Aside from the historic interest shown by the main actors of the international system in having the world's best specialists, what has made skilled personnel more important than ever are the estimations that predict the existence of a talent gap. The causes behind this talent gap are multiple and complex but can be reduced to two main causes: (i) increase of specialized demand and (ii) decreasing demographic trends.

(i) There is a significant increase on a global scale of the specialized demand of new job positions requiring new knowledge and techniques: 9 out of 10 of the most sought-after positions in 2012 did not exist in 2003 (O'Halloran, 2015); around 8 out of 10 children who are currently 8 years old will work in positions that do not exist nowadays, which means that only 21% of those children will work in positions that currently exist; it is also estimated that 47% of current positions might disappear in the next 20 years and 90% of those that remain will have to radically change (The Economist, 2014).

(ii) The demographic trends linked to the low birth rates and the aging of the population in places such as Japan, Korea, Israel, the United States, and the European Union are also worrying. In the 1950s, people over 80 years of age numbered only 14 million in the world; currently, there are more than 100 million, and current predictions point to this number increasing to 400 million in 2050. With fertility falling below the replacement level in all regions except Africa, experts predict rapidly rising dependency rates and a decline in the OECD labor workforce of approximately 600 million by 2050 (OECD, 2017), which will be particularly acute in North America, Europe, and Japan. Statistics point out that, in order to maintain economic growth, the United States needs more than 25 million new workers until the year 2030, and Western Europe needs a little over 45 million (WEF, 2011).

The scarcity in the offer of human capital and the increasing demand for skilled workers means thinking that, in the near future, it will be imperative for many countries to assure an increased labor supply of foreign workers, which will encourage a widening of the talent gap and transform the problem of mobility of skilled personnel into a central conflict between international actors.

Public policies

The relevance acquired by scientific knowledge, along with the prospects of a medium-term talent gap, has stimulated international actors to fight for this talent. Many experts consider the main causes of mobility to be not only internal difficulties in the countries of origin of skilled personnel but also by active policies in the host countries that stimulate the demand for global talent in STI (Pellegrino, 2001, 2004; Pellegrino, Bengochea, and Koolhaas, 2013; Albornoz, 2001; Bankinter, 2011; OST, 2016).

Public policies designed to attract international talent have precedents in the second half of the 20th century when the number of skilled immigrants admitted in their host countries started to increase. The paradigm is the United States, especially after 1965 when an immigration law was passed (The Significance of the 1965 Act) which, for the first time, established a mechanism of preference based on professional qualifications. Afterward, the quotas for qualified immigrants were increased by the Immigration Act of 1990 due to pressure from diverse business sectors. The open legislation of the United States regarding qualified migrants was progressively adopted by other countries such as Australia, Canada, and Japan. European countries, for their part, have established programs specially destined to recruit qualified resources. A report by the European Commission, with the suggestive title *Europe Needs More Scientists*, points out the following:

> The current facilities for producing trained researchers are geared to a much slower rate of growth than is currently envisaged. Clearly, these resources could not meet this demand unless substantially augmented and/or reformed. It seems unlikely, moreover, that the problem could be solved simply by "throwing a lot more money at it." Major structural changes will be required, at all levels, in the various national procedures by which researchers are educated, trained and recruited. (European Commission, 2004)

The competition for the dominion of knowledge and the talent gap in some key areas for the development of STI has led developed countries to support diverse recruitment mechanisms, such as international scholarships, the offer of better-paid jobs or preferential working conditions, with the objective of capturing this talent in any place of the planet. In addition to the States' investments in R&D and researchers, the different actors try to encourage the mobility and the recruitment of this specific collective through national policies that make migration laws more flexible for qualified personnel as a strategy to facilitate their entry and permanence. Canada, Australia, and Great Britain are good historical examples of this policy, which have since been joined by other countries as Brazil, China, Spain, and Singapore.

Currently, an increasing number of countries are getting involved in the global competition to attract human capital through the redesign of their immigration policies. The latest example has been the former president of the United States, Donald Trump, who although has asked for larger restrictions in the entry regime for skilled personnel (H1B visa), at the same time has proposed to eliminate the traditional lottery system and replace it with another program based on the merit and the qualifications of immigrants. In 2013, approximately 40% of the 172 member countries of the United Nations explicitly declared their interest in increasing their skilled population through the attraction of international migrants or to retain their native talent. This interest has doubled since 2005 when only 22% of countries expressed this same intention (Parsons, 2015).

What is being proposed by the majority of developed countries are public policies to attract and capture skilled workers through more flexible and talent-friendly immigration regimes that can help companies and economies benefit from the globalization of qualified resources, and investing more in R&D and education to attract foreign students interested in the fields of STEM (Gilman, 2010). In short, what the implementation of this type of incentive policy seeks is to respond to the projected future unmet demand in the most advanced countries where there is a clear deficit of skilled personnel and only the attraction of migrants from other countries will be able to close this gap.

Impact

The mobility of qualified personnel has generated a redistribution of the *global talent* that increasingly separates those countries that can make full use of their own resources, on top of attracting and absorbing external talent, and countries that are neither able to retain their own talent nor have policies for the repatriation of nationals. It is estimated that a third of scientists and engineers educated in less economically advanced countries left their countries to work in developed nations. Africa's case is the most dramatic, as it is estimated that there are more African scientists and engineers working in the United States than there are working in their home continent (García Guardilla, 2010). At the top of this hierarchy is the only country that maintains a positive balance with all others in terms of skilled migration balance, the United States; on the opposite place, we find countries in which intellectual institutions and industries are so weak that they cannot retain the majority of their talents, principally in Africa, Latin America, and some Asian countries (Meyer, Kaplan, and Charum, 2001).

This unequal redistribution of highly skilled personnel throughout the planet has been understood by experts in two different ways: On the one hand, those who analyze the phenomenon as *brain drain* by understanding that it greatly and directly affects the development of sender countries (which are less economically developed) because they cannot avoid losing their most qualified personnel while benefiting the recipient countries (which are the most developed) in their economies, the stimulation of production and the generation of STI. On the other hand, some consider skilled migration as a *brain gain* phenomenon by understanding the mobility of skilled personnel as a more complex topic, where qualified immigrants can bring their knowledge to their countries of origin, returning physically and also through new ICTs. In this sense, it is understood that qualified immigrants can create links and connections with global sources of knowledge, people, and funding and that they can share this knowledge with their place of origin through technology or, eventually, by returning and taking with them a high level of human and social capital. From this perspective, it is considered that there are benefits for all the actors involved through what is called the *global brain chain* (Faist, 2005).

The majority of international actors are interested in having highly skilled personnel as a main tool and motor in the generation of scientific knowledge and increase their relevance in the international system. At the same time,

States, companies, universities, and scientific institutions have started a hard battle to enlist the services of the most important *talents*. The upward global trend of bidding for talent continues strong as a result of a complex interaction between different actors: companies chasing scarce talent; developed countries' governments that try to attract these flows through public policies; less developed countries' governments unable to retain their own professionals; and, finally, qualified people that seek better labor options outside of their countries of origin to improve their socioeconomic conditions.

Intellectual property rights and patents

A second point of conflict in the field of ISR is not related to the highly skilled personnel that generates STI, but instead to the appropriation of scientific knowledge itself. It is one of the latest and most intense conflicts in the field of ISR and it is focused on the dispute about the way of managing, controlling, and appropriating knowledge.

The conflict is mainly caused by the relevance that STI has acquired in recent decades in the economic field as a source of competition between companies, which has meant the development and the boom of patent and private property rights systems applied to the results of scientific research activities. From its origins in 1978, the patent system has grown exponentially and has meant a conflict of interests between the *right to intellectual property* and what is called *public goods*. This private appropriation of knowledge, which has generated a strong controversy in the international community, is at the center of multiple international conflicts.

Origin

The growth of the means of scientific knowledge privatization was caused by several legal and governmental decisions taken in the United States that have had a great impact on the global system. Following Lander's (2008) analysis, we can differentiate two periods: (i) the evolution of the privatization of scientific knowledge within the United States and (ii) the international expansion of the process of scientific knowledge privatization.

(i) The start of the conflict goes back to the '80s in the United States when mercantile logic started to extend to the production of scientific knowledge. Lander describes it as a process of commercialization of scientific knowledge due to internal (universities and companies) and external (competition with other economies) pressures in the United States. "To this end, it was considered indispensable to eliminate all the legal and regulatory obstacles that made it difficult to strengthen the ties of these corporations with the extensive and vigorous university scientific and technological production system that the country had" (Lander, 2008). This university–company link is strengthened thanks to some decisions by the Supreme Court and the Patent Office, destined to increase the variety of things that can be patented, including diverse life-forms, genes, and even some therapeutic procedures. Among the main laws passed, we can mention the following: the Bayh–Dole Patent and Trademark Laws Amendment (1980) law, aiming to promote collaboration between commercial companies and nonprofit organizations, including universities, and the Federal Technology Transfer Act (1986), which authorized the commercialization of findings made in federal laboratories and the participation of scientists in the profits of companies that make use of these discoveries. The acceleration of this process resulted in a system of more commercialized production of scientific knowledge, in which companies had a larger access to the discoveries made in university laboratories, as well as the increase in the variety of products that can be appropriated by individuals and companies under the protection of the legal system.

(ii) The second step was to extend that same logic to the international level by strengthening protection systems (copyright and invention registration) to the detriment of the public domain. For that, an international legal and administrative architecture was built to watch over the interests of knowledge creators, who can temporarily make beneficial use of their discoveries. This began to appear after the revision of the World Intellectual Property Organization (WIPO) treaties, carried out in 1996, and the Agreement on Trade-Related Aspects of Intellectual Property Rights (TRIPS), negotiated in 1995, in the frame of the treaties by which the World Trade Organization was created, which eventually resulted in creators' interests taking precedence over users' interests, extending the exclusive rights and the temporary monopoly of intellectual property. As de Sousa Santos (2005; 2015, 2018, 2019) points out, what really happened was universalizing a radical expression of Anglo-Saxon liberal commercial law.

Vertiginous evolution

Ever since the strengthening of private protection, the number of patents has not stopped growing in the world order. As the WIPO points out, it took eighteen years (1978–96) to reach 250,000 patent applications, but in the next four years (1996–2000) the number doubled, and in only one year (2013) there were another 250, 000 new applications. On a global level, the number of international patent applications submitted by the Patent Cooperation Treaty

(PCT) registered an increase of 7.3% in 2016, which was the largest increase since 2011 and the seventh consecutive year of growth. With the exception of 2009, when the world financial crisis caused a drop, the submission of applications increased each year. It is estimated that there were 265,800 patent applications presented in 2019, which means almost 4 million international applications have been presented through the PCT system since it began operations in 1978 (WIPO, 2020).

With 58,990 PCT applications, applicants residing in China filed the most applications in 2019. This was the first year since the PCT System began its operation in 1978 that applicants from the United States moved down to the second place, with 57,840 PCT applications filed. They were followed by Japan, Germany, and the Republic of Korea (WIPO, 2020).

Clash of interests

Internationalization and intensification of this phenomenon have stimulated the debate on the private or public propriety of scientific knowledge, which suppose the clash of two rights protected by the Universal Declaration of Human Rights and the International Covenant on Economic, Social and Cultural Rights. This tension is evident from the two paragraphs (paragraphs 1 and 2) of Article 27 of the Universal Declaration of Human Rights: on the one hand, it is considered that "everyone has the right freely to participate in the cultural life of the community, to enjoy the arts and to share in scientific advancement and its benefits"; on the other hand, it is specified that "everyone has the right to the protection of the moral and material interests resulting from any scientific, literary or artistic production of which he is the author." This contraposition generates ambiguities and, thus, conflicts between the different actors that are trying to assert and maintain their rights and interests.

In the beginning, the patent system was established as an effort to try to mediate between those antagonistic rights, expressed in international legal systems. The origin of the conflict is the establishment of a system that, on the one hand, tries to protect the rights resulting from the production of scientific knowledge but, at the same time, seeks to contribute to the dissemination of the new knowledge. The goal of the protection systems was to encourage the growth of knowledge, as well as innovation, by setting a fixed time limit for the protection of intellectual property where only its author can benefit from the redistribution of their creation and, once this period was over, their proprietary rights would be terminated and the item would enter public domain for the benefit of all. From an economic point of view, the main objective is that intellectual property rights allow the creator to recoup the cost of the initial knowledge investment, having been given exclusive temporal rights, and encourage more creation (UNESCO, 2005).

However, in practice, a part of this purpose seems to have been distorted because, although it stimulates the production of new knowledge by some actors, it is also used to monopolize and control their benefits. As Baker (2012) points out, "An undeniable source of motivation, it has ushered a large number of noteworthy inventions. However, the perspective of benefiting from a monopoly also favors opportunistic behaviors, without any social value. The more important these gains are, the more this type of attitude spreads." The intellectual property system ended up becoming a sort of monopoly where inventors and creators are trying to extend the copyright period beyond the established limits; it has also become a strong weapon in the fights and conflicts between companies, as it has become common for patents to be used to fuel expensive legal battles. Companies such as Apple, Samsung, and Google are accustomed to mutually attacking each other for patent violations with each new product they put in the market, seeking to gain a threatening verdict to delay the commercialization of competing products for a few weeks, or even months, which is valuable time to seize a great part of the market (Baker, 2012).

In sum, intellectual property rights have become one of the most serious and complex issues in the current international agenda (due to the serious political, economic, and social consequences it can have) and have been one of the factors that has intensified the conflicts between the main actors within ISR.

4.3 Cooperation Dynamics

Historically, researchers, scientists, and institutions linked with science have had a strong reputation for working beyond the borders marked by the Nation-State, overcoming particular national interests to achieve common solutions for problems that, in the majority of cases, are shared by all of humanity. This internationalist logic, which is in the nature of the scientific enterprise, is a great opportunity to stimulate cooperative relations in a world that is increasingly moving towards accelerated processes of interconnection and intercommunication.

Throughout history, scientists have been in contact even during the tensest moments of a relationship and, many other times, they have been the first step in rebuilding from a conflictive link through the signing of scientific and technological exchange and cooperation agreements. For example, during a large part of the Cold War, scientific organizations in the United States and the Soviet Union fulfilled an important role in the agreements for the control of nuclear armament.[4]

Currently, distinguished specialists in STI (Fedoroff[5] and Bokova[6]) point out the emergence of an international context where new interrelation patterns, allowing to open a larger space for cooperation and joint development between STI actors. In an international system characterized by anarchy and conflict, there are increasingly more opportunities for cooperative links in the field of ISR. As Irina Bokova points out, "There are two possible scenarios for the way in which the geopolitics of science will shape the future. One is based on partnership and cooperation, and the other on efforts towards national supremacy. I am convinced that, more than ever, regional and international scientific cooperation is crucial to addressing the interrelated, complex and growing global challenges with which we are confronted" (UNESCO, 2010a).

In recent decades, cooperative relations between the actors in ISR have grown both in number and in intensity, and many new spaces or areas can easily be identified where cooperation is growing and spreading between actors linked to STI.

International cooperation

Bilateral, regional, intergovernmental, or transnational cooperative links between the different actors in the international system have been common practice between Nation-States and have extended with a growing strength in recent decades. This is because, aside from the strengthening of the national basis of innovation and knowledge, States began to consider scientific cooperation as "an effective agent to manage conflicts, improve global understanding, lay grounds for mutual respect and contribute to capacity-building in deprived world regions" (Flink and Schreiterer, 2010).

For that reason, the growth in bilateral and multilateral international scientific cooperation has become a common trait in all countries: the traditional State actors of scientific knowledge (the United States, the European Union, or Japan) show greater predisposition to collaborate with potential partners all around the planet; countries considered as emerging powers in STI (South

---

[4] A long list of mutual disarmament treaties and negotiations between Russia (and its predecessor the USSR) and the United States can be mentioned, including SALT I (1969–72), ABM Treaty (1972), SALT II (1972–79), INF Treaty (1987), and START I (1991).

[5] Irina Bokova (2010), Director-General of UNESCO (2009–17).

[6] Nina Fedoroff (2009, 2011), Science and Technology Advisor to the Secretary of State of the United States Government during the Barack Obama administration (2009–13).

Africa, Saudi Arabia, or Brazil) are greatly interested and predisposed in reaching cooperative agreements for the promotion of their own scientific and technological systems; and, lastly, a great number of developing countries are also trying to use cooperative links as a mechanism to boost their STI systems. This growing international cooperation (in all its forms) in the field of ISR is seen in the majority of proper scientific tasks that show a clear trend towards global cooperation: (i) more than 25% of global research projects are the direct result of international cooperation, having been written by authors from more than one country (UNESCO, 2015); (ii) the number of internationally co-authored papers has more than doubled since 1990; (iii) researchers have increased their international mobility, traveling to work with the best colleagues in their fields to have access to better resources and to share ideas and installations; and (iv) scientists are aided by cross-border funding from international organizations, multilateral initiatives between governments and research councils and multinational funding organizations (The Royal Society, 2011).

On an intergovernmental level, international cooperative links have also increased, and organizations such as UNESCO or the OECD have redoubled on their efforts in topics related to STI. UNESCO, as a dedicated science body, has intensified its work in policy advice and capacity building in the area of science and technology, especially in less developed countries; at the same time, it also faces global challenges that require the cooperation of the whole of the international scientific community, with topics such as social transformations, the environment, or sustainable economy. For its part, the OECD also carries out an extensive work agenda linked to cooperation and collaboration in the field of STI focused on its member countries. Among its tasks, we highlight the monitoring of statistical indicators and dimensions linked to STI and education, which function as references for the evaluation and feedback of educational and scientific policies of their member countries (such as the PISA report or reports on STI and industry).

Regional and interregional scientific cooperation projects have also experienced a great boost in recent decades due to the rise and success of many regional integration processes. There are notable examples of regions that are implementing institutional cooperation agreements so that scientists and researchers can share resources and knowledge, tackle sensitive and strategic topics, share physical resources, or simply share the same scientific language. These processes have been strengthened by strong political support, such as from the European Union, the African Union, the Association of Southeast Asian Nations, or the Southern Common Market, who develop their own research strategies and help coordinate scientific efforts in their regions and their larger spheres of influence (The Royal Society, 2011). A good example of this phenomenon is the case of the European Union: the strategy is known as the *Bologna Process* (1999) led to the creation of the European Higher Education Area, the European Research Area, and the European Commission's Framework Program, which is the main tool through which the European Union countries collectively make their investments in STI.

Lastly, *South–South Cooperation*, which links developing countries to build capacity and share knowledge in the southern hemisphere, has also started to grow. Among the most interesting examples we find the agreement between India, Brazil, and South Africa, which have joined forces to promote scientific cooperation through the *IBSA Initiative* and the *International Science, Technology and Innovation Centre for South-South Cooperation* project, established in 2008 by the Malaysian government, whose main goal is to transform it into a cooperation platform for matters related to STI for G77 countries (The Royal Society, 2011).

STI diplomacy

The increase in the use of *science, technology, and innovation diplomacy* (STI Diplomacy or STI Diplo) in recent years has led to the rise of a strong international cooperation mechanism between the international actors that use scientific knowledge as a catalyst for agreements. Together with traditional diplomacy, the development of STI diplomacy represents an interesting instrument for building international cooperation strategies concerning scientific and technological development.

Science diplomacy has evolved and now includes new actors, new strategies, and new approaches. For that reason, nowadays *STI Diplo* is considered a new type of diplomacy, which is an increasing phenomenon, broader, deeper, and more complex than traditional science diplomacy.

STI Diplo is defined as the use of scientific, technological, and academic collaborations among international actors to address common problems and to build constructive international partnerships (Federoff, 2009; Lijesevic, 2010; The Royal Society, 2010; Leijten, 2019; SciTech DiploHub, 2019). STI Diplo is used to address a common international agenda where many of the global challenges related to education, health, economic growth, energy, or environment need the collaboration of all the STI players in the international system (UNESCO, 2015; World Science Forum, 2019). The scientific community works beyond national boundaries on problems of common interest, and, therefore, these channels of scientific exchange can contribute to new cooperation relationships. In this sense, STI can be understood as a new source of *soft power* (Nye, 2010, 2017) or *smart power* (Flink and Schreiterer, 2010; Flink and Ruffin, 2019) that can be used as a constructive tool for diplomacy that is not only used by State actors (traditional diplomacy) but also by a wide range of non-State actors (cities, regions, NGOs, etc.).

Even though something similar to STI diplomacy has been practiced throughout a large part of the 20th century, the new global context has given new impetus to the development of this instrument of international cooperation between actors due to the combination of three factors:

(i) a new global framework of globalization and scientific and technological revolution, incorporating new dynamics, topics, and actors to the international agenda;
(ii) the emergence of a new *knowledge society*, where scientific knowledge is revalued and linked to political and economic power;
(iii) a new scenery in international politics that offers the possibility of building and using soft power as a new mechanism for foreign policy.

In this new international context, traditional diplomacy understood as military power together with political and economic coercion, coexists with new forms of negotiation. Scientific knowledge has played a leading role in the development of the capacities of *hard power*, such as military technology; however, now STI has also started to play an important role in diplomacy based on soft power. The scientific community has worked beyond national frontiers on shared issues, so it is well positioned to support new forms of diplomacy that need nontraditional alliances of nations, organizations, and sectors. If aligned with broader foreign policy objectives, these channels of scientific exchange can contribute to coalition building and conflict resolution. For example, starting a scientific cooperation process in topics such as nuclear nonproliferation or taking care of the environment can sometimes provide an effective path to reach other forms of political dialogue that allow agreement in other sensitive areas.

Currently, a country's soft power is, at least, as important as its hard power, and for this reason, the demand for STI diplomacy as a new foreign policy tool is increasing. Flink and Schreiterer (2010) point out that "as a matter of fact, more and more countries are up to incorporate [science and technology] in their diplomatic toolkits in order to buttress their [international relations] with ties to civil society, especially in those regions in which official relations are somewhat tainted." Soon after the end of the Cold War, many countries started to worry about their R&D performance and their international competitiveness, but only a few considered linking STI matters in their external affairs to be better prepared for what could be the future scenario of a war for talent and scientific resources on a global scale.

Unlike more traditional diplomacy carried out by states, the field of STI diplomacy is not only related to State actors, but also to a large number of varied non-State actors such as IGOs and companies. Epistemic communities, including national academies and scientific organizations, also play an important role in STI diplomacy, particularly when formal political relations are weak or tense, enabling new or different kinds of cooperative and associative interactions. The variety of actors involved in these activities must be extended to include NGOs, multilateral organizations, and other informal networks. Even scientific diasporas are strategically important because scientists abroad are often keen to maintain close relations with their own countries of origin, even when the relations with the host country are not ideal.

For many, in an international scenery dominated by STI, scientific knowledge will be increasingly used as a diplomatic resource and STI diplomacy will play a predominant role as a tool to design approach, negotiation, and cooperation strategies between actors in the international system as a whole (Yakushiji, 2009; Lijesevic, 2010; Flink and Schreiterer, 2010; Flink and Ruffin, 2019; Leijten, 2019; SciTech DiploHub, 2019).

Links between state, company, and university

The links between States, companies, and universities have become one of the cooperative relations that has intensified the most in recent years as a result of new interpretations that consider the production of knowledge and innovation a direct consequence of the interactions between these actors. It is not strange, then, that more and more countries and regions decide to establish public policies aiming to articulate and coordinate these three actors' tasks with the final objective of stimulating economic and social development.

These new interactions between States, companies, and universities are formalized under the new knowledge production model called *Triple Helix* (Etzkowitz and Leydesdorff, 2000; Leydesdorff, 2012), which is focused on the interaction between universities, industries, and governments as a key element in scientific production and economic growth. The relation generated between these three international actors is dynamic and has evolved in time both in quantity and intensity. Traditionally, these actors produced knowledge individually or with a weak collaboration; however, with time, this division disappeared, and they began to act jointly, each adjusting their role to the necessities required by this triangular cooperation:

– *Nation-States* have abandoned the idea that national or regional innovation systems on their own are the best way to generate STI and have chosen

to encourage links with universities and companies. In the majority of public plans and policies, there is a focus on socioeconomic contributions as a result of new linkages that must take place, strengthening the relation between the State, university, and company and connecting policies with the environmental needs.
- *Universities* are being put under strong external pressures to diversify their role and extend their links outside of the traditional academic sphere, as their role is seen as especially strategic in the new global context. This is the origin of the university's *third mission*, by which it must fulfill a new task in knowledge societies and economies, focused on contributing to the economic and social local development through the production of knowledge-based innovations. This function is in addition to two functions carried out by the traditional Humboldtian university: teaching and research.
- Last, *companies*, needing scientific knowledge and innovations for their own competitiveness, seek in universities and governments strategic partners that would allow them to generate developments they could not achieve otherwise. The fact that States plan their public policies including other non-State actors and universities provide cognitive resources for the creation of products and resources linked to innovation is a great opportunity for companies.

Each of the three actors contributes to the generation of links and interactions with the rest of the actors through their own resources: the *university* does so through its researchers, its new technology transfer offices, and the generation of incubators; the *Nation-State*, does so through adapting the legal and institutional framework, the regulation of property right, and giving funding and grants; finally, *companies* do so through the contribution of private investments and the generation of research, development, and innovation (R&D&I) activities. The result of these cooperations implies the interaction of multiple sectors of each of the three actors: academic researchers that become the businessmen of their own inventions; businessmen that work in university laboratories or transfer offices; university personnel or employees that manage national or regional agencies charged with technology transfer, among other examples.

The acceleration of the interactions between States, companies, and universities is an answer to the new model of producing scientific knowledge and innovation, which essentially seeks the generation of knowledge through the linkage of these three actors. The key is stimulating the complementary link and the cooperation between actors to develop more and better knowledge through a process that benefits all parts equally—a win–win game.

Interuniversity cooperation

The new global context has also allowed and encouraged cooperative links between universities throughout the world. The main principle behind international cooperation between universities is based on the collaboration and complementation of their capacities for the undertaking of joint activities and on the association with other university actors for the achievement of interests and mutual benefit. For UNESCO, this cooperation between universities is "a condition sine qua non for the quality and the efficiency of the operation of higher education institutions" (UNESCO, 2005).

The phenomenon of interuniversity cooperation alludes to the horizontal interaction and collaboration between universities. In mutual cooperation, universities have found an excellent mechanism to achieve mutual benefits in a variety of topics, such as (i) institutional organization by sharing criteria and establishing agreements about administrative management policies; (ii) institutional strengthening and projection on a national, regional, and international scale; (iii) the improvement of the undergraduate, postgraduate, continuous, and online educational offerings through joint programs; (iv) the education and specialization of researchers and scientific research processes; and (v) the increase in the extension and transfer of scientific and technological power to other social actors.

The cooperation model used by universities has changed alongside the international context, allowing the university to better adapt to the new demands to which it has been put under as one of the main actors in the production and transmission of scientific knowledge. As Sebastián (2004, 2010) points out, while collaboration activities had been formerly based on a *spontaneous* model, characterized by considering cooperation between universities as an external element, additional to classical institutional policies, universities nowadays are choosing an *integrated* model, which sees cooperation as a strategic element of the university. This new model conceives of cooperation as a key aspect of institutional development and of the university's internationalization process, now carried out through bilateral and multilateral cooperation agreements, such as networks or consortiums. "These instruments extend the benefits of cooperation by increasing the possibilities for interaction and collaboration" (Sebastián, 2004).

Interuniversity cooperation has become a strategy and an instrument of great value for universities if they are to survive the new dynamics of the 21st century. As Moreno Alegre and Albáizar Fernández (2008) point out, "universities are increasingly looking for strategic alliances and forms of interuniversity cooperation (networks, associations, consortiums, societies) that would give them a better capacity to attract and retaining talent (both students and professors), a more prominent international presence, a better use of resources and, in short, a greater ability to compete in an increasingly demanding national and international scene."

4.4 Competitive Interactions

The competitive dynamic has become another one of the most frequent and relevant interactions in the field of ISR. The recent changes conceived in the international system have consolidated a new role for scientific knowledge as a central resource in the global market for the generation of competitiveness between countries, companies, and institutions. It is a new world order where competition between actors acquires visibility and transcendence as a linking element and where the social and economic progress of a nation is increasingly dependent on its capacity to make good use of scientific resources, attracting talent and capital, and creating strategies to use their R&D more efficiently to ensure its competitive advantages (Flink and Schreiterer, 2010). Essentially, it is about understanding that STI has gained an important and growing role in the competitive dispute for market power and influence within the international system.

The core of competitive interactions is the economic value that knowledge has acquired in this new international context. Nowadays, societies are based on the generation and industry of knowledge, which is bought and sold, imported, and exported as any other good in the market, and that a monopoly in the production and distribution of knowledge no longer exists, which opens the competition to new providers. As Olivé (2005) points out, "we are witnessing a new race for knowledge, for the construction of systems suited to produce it, and for the conditions that give different people and social sectors the capacities to make use of it and apply it to the resolution of their problems and the development of their potential."

This competitive logic is what has pushed the main State and non-State actors in ISR to increase their expenditures and investments in R&D and avoid being excluded from the new STI global market.

Competitive race between States

The competition between States is considered a natural conduct in international relations, and especially if that competition is due to strategic resources. For this reason, for several years now, States have begun to compete decisively for the development of their STI systems through the design of complex public plans aimed at investing in R&D, developing infrastructures, and training highly skilled personnel.

A look at the evolution of States' investments in R&D and researchers over the last few years allows us to understand the interest scientific knowledge holds and how, in this way, a new competitive race has started over it. According to NSF (National Science Board and National Science Foundation, 2020), the total global R&D expenditures have risen substantially, expanding threefold between 2000 ($722 billion) and 2017 ($2.2 trillion). Among individual countries, the United States was the largest R&D performer in 2017, followed by China, whose R&D spending now exceeds that of the European Union. Japan (8%) and Germany (6%) are next, followed by South Korea (4%). France, India, the United Kingdom, Russia, Brazil, Taiwan, Italy, Canada, Spain, Turkey, and Australia account for 1%–3% each of the global total.

Observing this data allows us to understand the strong competitive race most countries have embarked on in order to optimize their STI sector. The traditional actors of scientific knowledge (the United States, the European Union, and Japan) are trying to preserve their historical privilege in the field of STI by carrying out investments of their Gross Domestic Product (GDP) in R&D. Indicators are showing that the investment, in all three cases, is very significant; globally speaking, however, their global hegemony is on the decline. Although the United States spent more on R&D than did any other country in 2017, its global share since 2000 fell as R&D spending rose in many Asian countries, especially China. Despite the United States continuing to spearhead innovation (including new emerging technologies), the competition in emergent markets has increased, which suggests a changing landscape in geopolitical and geoeconomic terms.

On the other hand, we can see China's prominent rise, accounting for almost one-third (32%) of the total global growth between 2000 and 2017 and having become one of the main actors in the production and generation of scientific knowledge at the beginning of the 21st century, overtaking Japan and the European Union in investment and closing to the United States in absolute numbers. China is closer to becoming the leading global R&D investor and is forecast to surpass total spending of the United States by no later than 2025 if current spending trends continue (R&D World, 2020).

Finally, there is a new group of State actors (India, Brazil, South Korea, Turkey, Russia, etc.) that, due to the rapid growth in their expenditures in R&D and researchers in recent years, are significantly reducing the traditional gap separating them from countries more advanced in STI matters.

Competition between companies

For companies, the competitive dynamics is the natural way of interacting with the rest of their peers and the normal method they use to achieve their economic goals. Precisely because of this, it is simple for them to competitively link themselves with other actors, as it is their natural logic: in essence, competition is what determines if a company grows, develops, and is successful, or if, on the contrary, it fails and disappears.

What is novel in the new dynamics of competition between companies is, fundamentally, the relevance that scientific knowledge has acquired as an indispensable resource to compete in the corporate sector. Little by little, knowledge has become a key factor of production to generate innovation, competitiveness, and growth. Currently, knowledge and creativity are considered determining factors for innovation, both to sustain and elevate competitiveness as well as to adequately react to the problems that might have been generated in the contextual variants (social, economic, cultural, political, historical). Precisely, it is for this reason that companies have made a strong bet for the investment in STI as a key strategy to achieve positive results. Amazon spent the most on R&D in 2018, with about 22.6 billion U.S. dollars. Alphabet, Volkswagen, Samsung, and Intel rounded out the top five companies with the highest R&D spending (Statista, 2020).

Currently, the key to competitiveness between companies lies in three main areas, all of which are related to scientific knowledge:

– *Innovation*: The key element of the new *knowledge economy* is product, process, and service innovation. For this reason, companies are looking for higher levels of competitiveness by making a large part of their companies' R&D&I investments in the field of intensive industrial research, in such sectors as pharmaceutical products, technology, electronics, and computer hardware. However, in the next years, the investment in developing software and complementary services to their traditional offer, with the goal of offering a better service and user experience, is going to be prioritized as much as producing new products (PwC, 2018).
– *Markets*: In the second place, we find the companies' search for higher levels of competitiveness in new international markets. In recent decades, transnational companies have started a strong competition to reach emerging economies, China's precious market in particular. There, many US and European companies are investing in R&D and employment due to the increasing offer of skilled workforce, the low salary costs, and the favorable incentives offered by local political institutions. In these fields, companies establish strategic links with local universities and build research centers in the new markets to which they get access. For example, Pfizer, a company with one of the highest investments in R&D (7.9 billion dollars in 2017), has developed its own R&D center in Shanghai, as well

as research alliances with the main Chinese universities. There are many cases of transnational companies that make such investments: Pfizer, Microsoft, Ford Motor, Boeing, Intel, Cisco, and IBM, among others.
- *Highly skilled personnel*: Lastly, the competition between companies is determined by the attraction of global talent to join business organizations. In a talent-gap scenario, the scarcity of highly qualified personnel stimulates the fierce competition between companies to enlist the best and brightest talents on a global scale and encourage companies to promote themselves as attractive workplaces by offering better work and professional conditions to qualified resources.

University rankings

Universities represent a fundamental resource in the new knowledge society, as they are a key piece as centers of training, creativity, innovation, entrepreneurship, transfer, and attraction of investments and talents. Precisely, in the face of this revision, it is not strange that universities have also adopted a competitive logic. Nowadays, universities have become global international actors, as they increasingly compete among themselves to attract funding, professors, and students, because the university's reputation is now built on an international level, which means competing with other institutions on a local, regional, and a global level. As UNESCO (2015) points out, universities have become true *trademarks* that compete for students, professors, researchers, sources of funding, strategic partners (both national and international), and corporate links.

The competition between universities has become institutionalized in the global market due to accreditation and evaluation systems, carried out by public and private agencies on different levels. The most visible aspect of competitive internationalization is the recent appearance of global university rankings, which represent one of the most evident effects of academic globalization.[7] These rankings explicitly show the competitiveness of a university by comparing it with its peers over a series of criteria which, normally, are linked to infrastructure, academic offering, teaching quality, research prestige, and scientific production in papers and patents, among other indicators. Currently,

---

[7] Despite the proliferation of many university rankings, the three most renowned and influential ones are (i) the Annual Academic Ranking of World Universities (ARWU), first published in June 2003 by the Institute of Higher Education of the Jiao Tong University of Shanghai (UJTS); (ii) the Academic Ranking of World Universities by the Times Higher Education Supplement (THES), London; and (iii) the Quacquarelli Symonds (QS) World University, published by the British company Quacquarelli Symonds.

and from the point of view of academic institutions, the concentration of knowledge is found, in most cases, in the so-called *world-class universities*, which are at the top of world rankings and which guarantee innovations and knowledge with impact on the world economy to global corporations (García Guardilla, 2010).

Even though all these rankings had been strongly criticized due to the methods they use, their influence and relevance have spread remarkably. The majority of national policies and plans on STI, as well as higher education, show as the main strategic objective to place their main university centers as high as possible in the international rankings list. Postgraduate scholarships offered by national agencies and international organizations require previous admission to one of the universities placed at the top of the university rankings. Currently, the bloc of Anglo-Western universities (United States, United Kingdom, Canada, and Australia) and the rest of Western Europe cover 93% of the first 100 places in these world rankings.

The extension and naturalization of the competitive logic between universities is a controversial phenomenon for many experts, showing discrepancies about the consequences of this open competition. For some intellectuals, the university's entry into commercial and competitive logic is a direct affront against the very nature of its activity and it can only be understood in the context of a process of the commercialization of scientific knowledge. Experts such as Boaventura de Sousa Santos (2005, 2015, 2018, 2019) point out the existence of neoliberal medium- and long-term projects that include different levels and forms of commercialization of universities, which are based on the following two main pillars: the decrease of State investment in public universities and the commercial globalization of universities. The end goal of this process is none other than the transnationalization of the university educational services market.

For other specialists, however, this competition is both necessary and beneficial to the university. Quintanilla (2007), for example, criticizes those understanding the university's competitive logic as an imposition by the capitalist system and considers that competition can bring out the best of the university, its scientists and its students for the development of STI: "Even if nowadays we would have a socialist instead of a capitalist economy, it would still be necessary for European universities to compete between themselves to be better and better, or to attract the best students and scientists, to guarantee the progress of science and technology, and consequently the increase in the population's well-being" (Quintanilla, 2007).

Fight for new emerging technologies

The so-called Fourth Industrial Revolution has pushed emerging technologies to the forefront of the international system as a new strategic resource for the actors' competitiveness. Artificial intelligence, robotics, quantum computing, synthetic biology, big data, and biotechnology are some of the new areas in STI that promise changes and applications that can radically alter the economic, financial, military, political, and social domains. These modifications will have strong geopolitical repercussions on the international system and will mean repositioning and new polarizations among the main actors. Conscious of this new reality, the majority of countries have begun a competitive race to develop these new emerging technologies in what has started to be called the new "Cold War for technology" (Eurasia Group, 2018).

In recent years, the majority of the international system's most important countries have designed national programs linked to investment and development in new emerging technologies applied to the promotion of *the digital economy*. Germany was one of the first in launching its program in 2011 (Industrie 4.0), followed by the United States, Italy, France, South Korea, Japan, China, India, Russia, Australia, Canada, and Spain. These projects represent strategic middle- and long-term visions that consider innovating in emerging technologies as a strategic resource for economic and social development and as a key element for a better geopolitical positioning in the future international system.

The emergence of new fields in STI that have a phenomenal potential for application in the economic sector has established a strong competitive race between the main countries of the international system. The United States maintains its scientific and technological leadership thanks to its still strong public investment (civilian and military) in R&D&I and its very dynamic private sector, one of the most innovating ecosystems in the world; however, the challenge by South-East Asian countries, mainly China, and the European Union's determination, led by Germany, forecast a very competitive future international system.

The United States continues to be the world leader in R&D&I with almost 30% of the global expenditure. The majority of its investments come from the private sector, which has always shown great dynamism and has turned into the best example of innovating ecosystems (San Francisco, Seattle, Boston, etc.). Public investment is also elevated, though it suffered cuts after the economic crisis of 2011. In recent years, the American government has tried to leverage the innovating capacity of its private sector by promoting many joint public–private initiatives with the objective of finding synergies between both sectors in the field of emerging technologies. The *Defense Innovation Experimental Unit*, dependent on the United States Department of Defense, was created in 2015 as a center of attraction for the private sector's innovations to promote the application of new emerging technology to military developments.[8] Located in Silicon Valley (a Mecca of innovation), its main objective is to change the bureaucratic culture of the State through the association with companies' and technological centers' innovative culture (Fojon, 2018).

China shows open interest in the leadership of the international system with strategic moves in the field of new emerging technologies. In June 2015, it launched its industrial program named *Made in China* 2025, where it openly expresses the vision of leading in a near future ten industrial sectors linked with high technology, including areas such as robotics, artificial intelligence, autonomous vehicles, and biomedicine, among others. The main goal is to reduce the dependency on key foreign technology by 40% in 2020 and making the Chinese economy *completely intelligent* by 2030. For this, China made an extraordinary R&D investment in researchers, infrastructure, higher education, highly qualified personnel attraction policies, and strategic investments abroad. As a result, international patents have increased by an annual 17.4% since 2007; scientific publications have grown by 22.9% per year, and graduates in hard sciences, which reached 6,081,600 in 2012, have increased by an annual 16.4% (Castro, 2018). At the same time, it is working with the objective of perfecting e-commerce, online banking, social media, and autonomous vehicle algorithms to dominate these technologies soon. As a consequence of all this effort, China has made very significant advances in the fields of quantum computing, synthetic biology, and artificial intelligence. Currently, China is the second ecosystem in the field of artificial intelligence, only surpassed by the United States, where more than 700 companies registered 16,000 patents in a single year (Campanario, 2017).

The European Union is the third relevant actor in this competition created by emerging technologies. Germany was the first government to promote a long-term strategy for the development of new emerging technologies applied to their industry in 2011 (Industrie 4.0), which was later copied by other countries. In addition, On February 19, 2020, the European Commission presented its digital strategy, which includes a data strategy and a White Paper on artificial intelligence. The three key objectives of that strategy are "technology that works for people," "a fair and competitive economy," and "an open, democratic, and sustainable society" (Rathenau Instituut, 2020). The high expenditures in R&D&I, researchers, publications, and patents show that

---

[8] *Project Maven*, launched in 2017 and linked to the use of artificial intelligence in the US military systems, is one of the most relevant programs in a new public–private association for the development of emerging technologies in the United States.

the European Union has decided to fight in the future scenario of ISR: the European Union's investment in the emerging technology sector has quintupled between 2011 and 2017, reaching over 11 billion euros (Ortega, 2018).

These are not the only states interested in the future of new technologies: countries such as India, South Korea, Japan, Australia, Canada, and Russia are also pushing political strategies and strong investments in emerging technologies with the objective of not falling behind in this global competition.

Lastly, in this equation, we cannot forget large technology companies such as Google, Facebook, and Amazon, which have acquired a global scale, weight, power, and influence similar to that of many States in the international system and which, in many cases, have very dissimilar interests to international powers such as the United States, China, or the European Union (Campanario, 2017), which opens a very possible scenario of tension and conflict between State and non-State actors.

In conclusion, it is evident that STI will have growing importance as a strategic resource, and the competition for having said resources will cause all actors to fight for obtaining them, and the result of this competition will largely depend on the new configuration of the 21st century's international system. For that same reason, UNESCO (2005) sees that "in the near future, knowledge will be the object of an increasingly contested competition between the many actors, which will profoundly influence and modify the geopolitics of the 21st century."

4.5 Asymmetric Relations

The interactions that establish links of hegemony and subordination between actors can also be considered frequent relations in the subfield of ISR. Historically, the main actors of the international system have established these types of asymmetric relations by projecting their power through a combination of persuasive and coercive actions, aiming to impose and maintain their particular and general interests. The differences in capacity, resources, and power between international actors have established an unequal global system where there exist types of relations where some actors prevail over others.

Historic inequalities

Asymmetric relations generating links of hegemony and subordination in ISR have been a constant throughout history and go back to the realization that it is a precious resource for which many different actors would fight. As a result, the configuration of ISR has been many times characterized by relations of one actor having hegemony over others.

In the historic evolution of scientific knowledge, the West, geographically understood as Western European and North American countries, has led the creation, dissemination, and reproduction of scientific knowledge. From its expansion during the 15th century, Western Europe inserted itself in a process that positioned it as the main *geocultural entity* of the world (Wallerstein, 1991), which was based on the idea of *modernity*. Throughout history, the West has occupied a central and hegemonic role in the international structure of scientific knowledge, built on the base of positioning itself as the essential motor of progress, universal civilization, modernity, and development. As David Slater (2008) points out, this self-perception is built on three interconnected ideas:

(i) The consideration of the West as possessing a series of attributes that are inextricable to it, such as rationality, democracy, modernity, and human rights.
(ii) The interpretation that these attributes are a part of European development and not a product of a cultural and historical admixture.
(iii) The idea that the development of these exclusively Western attributes is a transcendental step for all of humanity.

For the tradition of Western thought, the world is objectively knowable and can be captured by concepts and representations built on reason. From this radical separation between subject and object, rooting from Christian thought, secularized by Descartes, the construction of knowledge is objectivized and can, as such, be generalized and transformed into universal. This self-perception has a negative image of the non-Western culture. On this basis, modern Western thought was built, and it became the main and hegemonic axis in the world order. The West thus ended up positioning itself in a place of privilege, thanks to the fact that European countries, at first, and later through the leadership of the United States, were the first in the development of universities, prestigious centers of studies, and research institutes that they used for the benefit of their own cognitive, social, cultural, and economic development. This has allowed them to reach almost a monopoly of scientific knowledge, to the detriment of other zones and geographic spaces.

Power and knowledge

Critical theory and postmodernism considered that, throughout history, knowledge has been associated with power and sociopolitical, geopolitical, and geoeconomic domination schemes. With the advent of modernity and the rise of scientific knowledge, controlling this resource of power has become fundamental for all actors in the international system that wanted some primacy or hegemony in the world order. It is a thesis defended by thinkers such as Habermas (1968) or Marcuse (1964), who understand that sciences and techniques are the instances that assume the legitimating function of domination. From Ancient Greece, through the Roman Empire, medieval universities, 16th- to 19th-century European empires to the current American hegemony in the field of higher education, the control of knowledge has been one of the main attributes in the power accumulation scheme.

In this line of thought, García Guadilla (1996) points out that "traditionally, science and technology have been central elements for the legitimation of social systems and, for that same reason, scientific knowledge cannot be only thought of as an altruistic element, but rather must be understood as a substantial tool to understand the power of people, institutions or countries." Based on this relationship between power, knowledge production, and geographical space, the configuration of the geopolitics and geoeconomic of scientific knowledge has been historically associated with political, economic, and social structures of hegemony and domination; and essentially, what now comes into play is the "special relation between knowledge and power or, more specifically, between scientific knowledge and political and economic power, in very concrete and differentiated geographical spheres" (Ferrero and Filibi López, 2006).

Cognitive divide and dependency

Within the global scenario of the rise of STI as a central and strategic resource for the future agenda of global issues, the links of dependence between actors take on special relevance. A global analysis of the main dimensions linked to scientific knowledge allows us to see the clear hegemony exerted by some actors over scientific development and, as a consequence, the subordination into which many actors are forced.

Currently, five actors (United States, China, European Union, Japan, and Russia) jointly represent 70% of the world's researchers, 80% of the R&D expenditure, 88% of scientific publications, 99% of patents given by the United States Patent and Trademark Office; attract 8 in 10 higher learning international students; and house 99% of the world's top universities. In contrast, the world's poorest 50 countries represent 0.3% of the world's R&D expenditure, 0.8% of researchers, 0.6% of scientific publications, and 5% of the Triadic[9] patents (UNESCO, 2015).

---

[9] Triadic patent is the name given to an invention patented by the same creator in the three most important patent offices on a global level: the European Patent Office (EPO), the United States Patent and Trademark Office (USPTO), and the Japan Patent Office (JPO).

Statistics confirm the existence of completely asymmetric relations in the field of ISR. Some of the data is so overwhelming, and the inequalities between international actors so deep, that UNESCO has started to denounce the dangers of creating a dissociated global society where knowledge is distributed in such an unequal way and advocate for the creation of a knowledge society that would be a source of development for everyone on a global level. For UNESCO, hegemony and superiority relations are a clear threat for deepening the existing *knowledge gap* and substituting it with an acute phenomenon of *knowledge dependency* (UNESCO, 2005).

This same reality is seen with attention by numerous experts that worry about the existence of these asymmetric relations, the consolidation of some international actors' supremacy on scientific knowledge, and the serious repercussions of the unequal distribution of STI in the international system. The technological revolution and the expansion of the global economy are contributing to the generation of new dispersion and centralization dynamics that are forming what many specialists consider to be *new geography of centrality*. These new centers and peripheries recreate old differences and generate new ones, typical of the dynamic of hegemonic linking that is imposed between actors, which implies that scientific knowledge is spread in very specific centers (nodes), which generate larger inequalities between countries and regions (Innerarity, 2011, 2013; UNESCO 2015; Sassen, 2017; Van Dijk, 2020).

Chapter 5

PROCESSES

One of the most relevant topics in the field of ISR is the ensemble of operative processes that take place within it. As with any other system, ISR possesses specific methods of interconnection and interaction between its actors that generate particular internal mechanisms. In our current global context, it is possible to recognize some essential internal processes typical of scientific knowledge, which are a part of the operative mechanisms developed in the field of ISR but which, due to their relevance, influence the international system as a whole. Essentially, these processes describe and explain knowledge's life cycle from its creation to its final application and global governance.

It must be considered that all these internal processes of the ISR are part of a dynamics that is not linear, uniform, or progressive, but, on the contrary, these processes are strongly interlinked, which generates a more discontinuous, interrelated, and complex process. In practice, they are interactive processes that admit a certain degree of overlapping in time and whose protagonists intervene in diverse moments and play diverse roles. In reality, they are interactive processes where the interdependency between their mechanisms is the main characteristic and where each one of them is complex in itself due to the many actors establishing multiple interactions and carrying out many tasks with dissimilar results.[1]

Consequently, the goal of this chapter is to tackle and examine the almost indecipherable *black box* that scientific knowledge represents, studying each of the processes and mechanisms that form it in depth. For that, it is necessary to make a detailed and deep analysis of the five main processes in ISR: *production, intermediation, distribution, application,* and *governance*.[2]

---

[1] Each process must be understood as a receptor of *demands* generated by other processes and, at the same time, must also be thought of as *solution* generators that create *feedback* on other processes and are used for the regeneration of new scientific knowledge. Then, it is within pluri- and multicausal phenomena where all processes are linked and mutually influence each other, causing feedback loops.

[2] The OECD distinguishes the processes of *production, mediation,* and *use and application* as the three main instances in the process of creating scientific knowledge. These processes

5.1 Production

The *production* of scientific knowledge is one of the first processes that must be studied in the field of STI. The first approach allows us to see that the conjunction of the changes that have taken place in the global configuration of the international system, and the specific modifications that have happened in ISR (new actors, the redefinition of interactions, and the revaluing of scientific knowledge) have strongly affected how new scientific knowledge is produced. Specialized international organizations and experts coincide in the diagnosis of *a new age* in the process of knowledge creation (Gibbons et al., 1994; Nowotny, Scott, and Gibbons, 2003, 2006; Etzkowitz and Leyersdorf, 2000; Funtowicz and Ravetz, 2003; Carayannis and Campbell, 2006, 2009; UNESCO, 2015; Carayannis, Campbell, and Rehman, 2016; OECD, 2017). These changes involve new actors taking part, new kind of interaction dynamics, the reconfiguration of the main internal processes, the acceleration of production mechanisms, and the creation of new types of knowledge and new places where it can be produced. It is a true revolution, with yet unforeseeable consequences.

For an in-depth study of the new methods of knowledge generation emerging in the international system, one must, essentially, concentrate the analysis on two substantial changes in the current process of scientific knowledge production: (i) it is necessary to identify the main changes that are occurring in the methods of scientific knowledge creation, and (ii) it is necessary to observe the process of acceleration and intensification of the factors of knowledge production.

Methods of production

The first element to be analyzed in the production of scientific knowledge is directly related to the methods used for its creation. The idea of a production method hearkens back to the creation of science itself, considering the essential characteristics presented by the whole of the scientific research cycle. Some of the key elements of this mechanism are as follows: the actors involved (university, state, company, etc.), the interactions generated (two- or three-way links, or with even more participants), the type of knowledge created (basic or applied), the places where it is produced (inside or outside academia), and the validation and evaluation of the knowledge (expert or general).

---

are further subdivided in seven specific processes: "production-validation-codification-dissemination (which includes diffusion and transference)-adoption-implementation-institutionalization" (OECD, 2000).

The methods of scientific knowledge production are not uniform or homogeneous; on the contrary, they must be considered as dynamic and heterogeneous. They have evolved throughout history, which implies significant changes in the way new knowledge is created: from individual and almost solitary efforts of many scientists, through the traditional linear conceptions of progress in science and technology coming from the period after the world wars to the National Innovation Systems, the production of knowledge has always been evolving. Currently, it seems we are going through a historical transition cycle to new methods of knowledge production, where the process that generates science is substantially transformed in the majority of its essential parameters.

Traditional methods

Before the boom of *Big Science* in the middle of the 20th century, it was impossible to find a planned and articulate method of scientific knowledge production with State support and clearly delineated goals. Scientific generation was mainly carried out in a fragmented way and focused on universities and some sectors such as the military. The advent of Big Science after World War II meant a radical change in the scientific knowledge production methods, with the implementation, for the first time, of public policies stimulating a systematic method for the creation of science and technology. This *linear model*, based on an optimistic and deterministic view of STI, prompted large State investments with the objective of reaching economic and social development.

In the '70s, this view started to be revised as a consequence of the strong critiques it received from the sectors most skeptical about the role of science and technology, who discussed the need for a more complex approach to the phenomenon. From this context emerged the model of *Innovation Systems* or "*National Innovation Systems*" (NIS),[3] which tried to offer solutions to the theoretical criticisms and the practical transformation that started to manifest in industrialized countries. Through this model, innovation started to appear in political speeches about science and technology that before only talked about R&D; that was how the "I" of innovation was added to R&D (R&D&I) and public funding for science and technology started to be distributed also taking in account the results of research (González de la Fe, 2009). A large part of this model's success stemmed from its capacity to overcome the classical linear conceptions of the progress of science and technology, assuming the need for

---

[3] The model was first introduced by Bengt-Åke Lundvall (1985) as "Innovation Systems" and later by Christopher Freeman (1995) as "National Innovation systems."

the Nation-State playing a leading role but, at the same time, sharing this role with other non-State actors.

The successful application of Innovation Systems in many countries opened an interesting debate about the possibility of applying this production method to any other geographical area. Normally, the development of this model has been associated with developed countries, even though it is considered to be applicable to other nations through adaptations and adjustments. The successful cases of Finland or South Korea have led people to believe that, with constant policies applied throughout a large period of time, similar results can be achieved. Despite this, some critical stances point out that the success of an Innovation System is more dependent on external factors (such as the country's integration in the world economy or the international juridical context) rather than on the model itself.

New methods of production

The changes in the international system and the global reevaluation of STI have boosted new methods of scientific knowledge production that seek to go beyond the traditional approaches. Gibbons et al. (1994) and Nowotny, Scott, and Gibbons (2003, 2006) understand that these changes are precisely those who are promoting the emergence of a new method of creating knowledge (Mode 2) that substitutes the old and traditional way (Mode 1). In reality, it is a transitory (and coexistent) period between two methods of generating scientific knowledge, where the new method evolves on top of the traditional model. These changes not only affect the type of knowledge produced but also how it is produced, the context in which it is created, the way in which it is organized, the reward system used, and the mechanisms that control the quality of the products:

– *Places*: There are new places for the production and practice of scientific research. The new way of producing knowledge is not concentrating in universities and research centers; the environments that can potentially generate knowledge, the organizational methods, and the work methods have multiplied. The number of graduates and experts has grown over the classical disciplinary structures, and many of these new specialists have gone on to work in government laboratories, in industry, and consulting teams; as a consequence, the number of places where a competent research project can be carried out has grown substantially (Gibbons et al., 1994, Nowotny, Scott, and Gibbons, 2003, 2006).
– *Levels*: Carayannis and Campbell (2009) introduced the concept of Mode 3 to explain the idea of producing new scientific knowledge in different

levels: (i) at the individual or micro-level (entrepreneurs), (ii) at the organizational or meso-level (innovation networks), and (iii) the systemic or macro-level (democratic or capitalist system).
- *Actors*: A second change is the relevance acquired by some international actors in the production of knowledge, due to the significant increase in the offer and demand for scientific knowledge. On the one hand, institutions of higher education, and especially universities, have collaborated to increase the number of people, institutions, and actors that can generate scientific knowledge beyond traditional actors. On the other hand, the new international context has produced an expansion of the demand for all types of specialized knowledge, which is starting to be satisfied by nontraditional actors in the field of STI. The result has been the proliferation of new State actors (China, Brazil, South Korea, etc.) and non-State actors (think tanks, diasporas, epistemic communities, etc.) interested in playing a role in the processes of scientific knowledge production.
- *Interactions*: We also highlight the increase and diversification of interactions between the groups charged with the production of scientific knowledge. The new international context has promoted interactions thanks to the development of ICTs, which has set the scene for an explosion in the number of interconnections. The result is the establishment of a system of knowledge production that forms a network whose nodes extend all around the globe and whose connectivity grows continually. The expansion of the number, nature, and the reach of interactions between different places of knowledge production leads not only to more knowledge being produced but also to have more expertise on different topics.[4]
- *Application*: Another important change is the strong trend of producing new knowledge aiming to satisfy a larger number of demands coming from multiple sectors, especially from the economic sector. Essentially, praxis here is above any other consideration, by which an attitude of permeability to external exigencies and pressures is maintained, which enlarges the agenda of possible research topics.
- *Social responsibility and reflexivity:* At the same time that the production of knowledge is oriented to the application, there are also new demands to generate knowledge that correctly tackles new problems of a social nature. This implies the emergence of a new method of producing knowledge

---

[4] An example of this is the Triple Helix model (Etzkowitz and Leydesdorff, 2000), which proposes the creation of new knowledge through the three main actors in ISR: the university, the company, and the State. These three spheres, which previously worked independently, now have to work jointly through two- and three-way cooperative relations.

that can face the complexities, challenges, and uncertainties brought on by the new international agenda.
- *Transdisciplinarity*: There is an increased production of knowledge that goes beyond traditional disciplinary structures, typical of modern science, which has led to changes in the scientific agenda, the way of using resources, the methods by which research is organized, and how results are communicated and evaluated. García Guadilla (1996, 2005, 2010) points out that we are going through a revolution where the old production structures, superspecialized and based in atomized forms, are giving way to new ways of creating and generating knowledge. It is a return to the form in which knowledge was conceived in the Renaissance, but with the difference that now there is the possibility of creating structures that are able to concentrate large quantities of knowledge in integrated ways through interdisciplinary methodologies.
- *Quality control*: In the process of evaluation and control, pragmatic criteria are incorporated and new social actors are getting involved. It is what Funtowicz and Ravetz (2003) call the "extended peer community," which must be understood as the inclusion of other *legitimate participants* of public life in the production of knowledge. The involvement of a greater number of new actors means the completion of a triple objective: (i) increasing democratic participation, (ii) assuring the quality of the process and the final product, and (iii) achieving a wider distribution of responsibilities.
- *Methodology*: Great changes are also observed in the process of scientific research, with the apparition of new methods, techniques, procedures, ways to collect and analyze data, and different ways of managing and transferring new knowledge, at the same time that new methodologies from outside the strictly scientific world are included.
- *Results*: The generation of new knowledge is also facing the new challenge of providing adequate and definitive solutions in a more complex and uncertain international context. Funtowicz and Ravetz (2003)[5] point out the existence of a new historical moment where scientific practice must face the challenges caused by the collision of a plurality of actors, interests, values, and interactions and, at the same time, there is a need to create knowledge and make decisions, even before having scientific proof, knowing that the potential impact of these decisions will be very large. In this scenario, where risks are not quantifiable or the potential damage can be irreversible, it is impossible to go back to traditional methods of

---

[5] The *"post-normal science"* model is a concept developed by Funtowicz and Ravetz in 1993, which tries to explain how the new contextual conditions of the international system are affecting the methods of producing scientific knowledge.

producing scientific knowledge to solve a problem or to guide a public policy.

In conclusion, these new methods of producing knowledge are evolving and transitioning[6] and show that the changes that have taken place in the international system and within the ISR are especially affecting the way in which scientific knowledge is produced.

Alternative methods

From remote geographical areas (sometimes considered as peripheral) and from ideologies far from the mainstream, a very wide range of critical viewpoints have been developed. Those positions challenge many of the current fundamentals and, at the same time, offer alternative approaches for the understanding of the production of scientific knowledge. Essentially, criticisms accumulate around two main axes: firstly, the majority of production methods are based on the experience of developed countries and do not have a clear transference or adaptation to dissimilar geographical and/or economic spaces; secondly, it is criticized that many of the production methods frequently have a heavy commercial emphasis, pushing social and human development to the background.

Many of those critical theories intend to develop knowledge production methods that better fit the reality of their own context. Because of that, there have appeared *alternative models* that focus on methods of scientific knowledge production that essentially prioritize social aspects. Among these models, we highlight the one proposed by Núñez Jover and Castro Sánchez (2005), namely the *higher-education–knowledge–science–technology–innovation–society complex*, where the expansion and consolidation of a scientific knowledge-producing sector, which coordinates the efforts of different agents, is encouraged. It is considered fundamental for developing countries to create effective processes for the creation, distribution, and application of knowledge and connecting these new capacities with effective economic, social, cultural, political, and educational development strategies.

---

[6] For Gibbons, being in the initial stage of development, some of the practices associated with the new method are already creating a pressure that tends towards a radical change in the traditional science institutions, particularly universities and national research councils. It is not surprising, then, that some of these institutions are particularly resistant to such changes, which seem to threaten the same structures and processes that have been created to protect the integrity of the scientific corporation (Gibbons et al. 1994, 2003, 2006).

The key to this model is linking the production of scientific knowledge to social development efforts. In essence, it is a method of knowledge production with a *social focus* that is different from the *traditional* model and the new *economic-corporate* models. To go further in this model, Núñez Jover introduces two core concepts: first, *sustainable social development based on knowledge*, where the strategic connection between knowledge and development is established; second, the concept of *higher-education–knowledge–science–technology–innovation–society complex*, which allows articulating a model between the university and society that pursues social goals (Núñez Jover et al., 2020).

The main traits of the *social development* production method (Núñez Jover and Castro Sánchez, 2005) can be summarized as follows: (i) the participation of the State and other social actors in knowledge generation, (ii) a strong interaction between the university and society, (iii) higher education institutions as central actors in the social production of knowledge which should serve society, (iv) the appropriation and social pertinence of knowledge production, (v) an agenda centered on social needs, (vi) promotion of advanced and permanent education, (vii) knowledge geared towards sustainable development, and (viii) knowledge connected with the productive sector, communities, and other social institutions.

The majority of the defenders of these alternative models understand that the production method of *society-oriented knowledge* is not necessarily at odds with the academic rigor that appears in the *traditional academic model* or with the articulation towards the corporate world accentuated by the *economic-corporate model*. Although they understand that "focusing on the links with large corporations, producing papers or solving social issues" are not the same, they also consider that it is possible to develop a production of scientific knowledge focused on "the urgencies of social development, which are carried out with the academic rigor guaranteeing its quality and, at the same time, promoting the participation in wealth creation" (Núñez Jover, 2006). The goal is to order the objectives of knowledge creation in a different way, putting social development as the highest priority and all other objectives (traditional science, or science oriented to the productive sector) under the umbrella of social interests.

Increase in the factors of production

The second element to be analyzed with regards to the process of scientific knowledge generation is linked to the acceleration and intensification of the factors of science production. The creation of knowledge is the result of a long process of scientific research in which many factors are combined, allowing us to reach a specific production. These factors are boosted and funded by international actors that now are interested in producing knowledge through an extensive and complicated mechanism requiring large investments.

A statistical review of the set of factors linked with the generation of scientific knowledge in the field of ISR allows us to understand the interests, objectives, and behavior of the actors towards STI. Among the most relevant dimensions to be studied to understand the behavior of the main actors towards the factors of scientific knowledge production, we find (i) expenditures in R&D, (ii) expenditures in researchers, (iii) investment in higher education, (iv) development of research infrastructure, and (v) scientific papers and patents.

Research and development (R&D)[7]

The main intergovernmental organizations (UNESCO, the OECD, or the World Bank), research centers (The Royal Society, Battelle, AAAS, L'Observatoire des Sciences et des Techniques, Science-Metrix, Goldman Sachs, etc.) and experts on the topic (Arnow, Stirner, or Godin) consider the R&D indicator as best suited to assess and evaluate a country's efforts with regards to the production of scientific knowledge. Essentially, this indicator gives a precise idea of the intensity of a country's research and its capacity to invest financial and human resources in STI activities. In addition, R&D plays a central role in advanced economies in areas such as economic growth and job creation, industrial competitiveness, national security, energy, agriculture, transportation, public health and well-being, environmental protection, and expanding the frontiers of human knowledge understanding (CRS, 2020).

In this sense, the R&D expenditure must be seen not only as an economic variable but mainly as a political option. In a way, R&D expenditures show a country's attitude towards scientific production and allow us to identify the disparities with regards to the investment between different actors that might have similar characteristics. Political goodwill and civil society's commitment are key elements of a good system of knowledge and innovation production (UNESCO, 2015 and NSF, 2018).

The great increase in R&D expenditures in the last few years on a global scale confirms the main actors' interest in STI as a strategic resource of the

---

[7] It is defined as follows: "Total internal expenditure in research and development in a national territory during a certain period" (OECD, 2003); "total expenditure (current and capital) in R&D by all national companies, research institutions, university and government laboratories in a territory. It excludes R&D expenses financed by national companies but carried out aboard" (OECD, 2008); and "total current and capital (public and private) expenditures in a creative work carried out systematically to increase knowledge, including knowledge on humanity, culture and society" (UNESCO, 2009).

international system. Since 2000, total global R&D expenditures have more than tripled in current dollars, from $676 billion to $2 trillion in 2018 (NSF, 2018; CRS, 2020). After the incredible 45% increase in 2002–7, investments grew by over 30% between 2007 and 2013, to reach 2 trillion dollars in 2018, doubling the trillion from 2005 and the 772 billion from 2000 (NSF, 2018). There is a significant increase in the annual growth in national R&D spending between 2000 and 2017 in most countries. A few examples include China (17.3%), South Korea (9.8%), India (8%), Germany (5.4%), European Union (5.1%), the United States (4.3%), the United Kingdom and France (4%), and Japan (3.3%).

Researchers[8]

The second factor of scientific knowledge production is researchers themselves, who are in charge of generating knowledge. The relevance assigned to highly skilled personnel is increasing, both in quantity as well as in quality, as they are the only resources that can invent and spread knowledge. The quality of a country's human capital will be a differentiating element to increase national competitiveness in the international context, allowing a country to gain ranks in the new global power map; for that reason, there is a growing interest in investing in them and recruiting the brightest students and professors.

Proof of this interest in qualified human resources is the significant increase in the number of researchers on an international level, whose global stock has grown from 6,400,900 in 2002 to 7,758,900 in 2014 (more than a 20% increase in 12 years), and more than 9 million in 2017 (World Bank, n.d.). At the same time, the number of researchers per million people on a global level also grew from 959 (in 2007) to 1411 (in 2015) (World Bank, 2020).

Higher education

Alongside expenditures in R&D and researchers, investment in higher education is another of the most important factors of production linked with scientific knowledge generation. This social institution has the complex and vast task of educating, training, and qualifying the human resources that will later be tasked with producing new knowledge. As it was pointed in the World Conference on Higher Education of 2009, "at no time in history has it been more important to invest in higher education as a major force in building an

---

[8] The OECD's Frascati Manual considers the two main factors to give a dimension to the investment or expenditure in science and technology to be (i) financial resources invested in R&D and (ii) the human resources linked to this activity (OECD, 2003).

inclusive and diverse knowledge society and to advance research, innovation and creativity" (UNESCO, 2009).

The growth of higher education as an essential factor for the production of scientific knowledge can be observed in the continuous increase in the main indicators linked to it. This relevance can manifest in, at least, three specific fields:

– *The massification of higher education*: The massification of higher education is a strong process throughout the planet. Approximately 35% of the OECD's population has a university degree, and in countries such as the United States, the number increases to 47% (OECD, 2016; Lumina Foundation, 2018). Between 1970 and 1990, the number of students enrolled in higher education was more than doubled, rising from 28 million to 69 million. This phenomenon increased even further since the end of the 1990s, with a spectacular growth in the number of universities and student enrollment all around the world. The university population went from 102 million in 1999 to 153 million in 2007 and reached a record-breaking 200 million in 2017 (UNESCO, 2017). This progression is not exclusive of rich countries, but a booming phenomenon in Africa, Asia, and Latin America, as a consequence of the strong demographic growth.

– *Increase in higher education spending*: The relevance of higher education can also be appreciated in the determined efforts by a large majority of countries to increase their expenditures and investments in this field. Generally speaking, all countries show an increase in their higher education spending to some extent. The OECD countries spend an average of 5.2% of their GDP on education, of which one-third goes exclusively to higher education (OECD, 2016).

– *Increase in the number of doctors and researchers*: In the past 20 years, we can see an increase in the number of international students studying in universities worldwide and the increase in the number of doctors and researchers coming out of those institutions and inserting in the workforce to carry out functions linked with the creation of new scientific knowledge. Between 1981 and 1999, the number of researchers has increased by 127%, which represents a 7% average increase per year (Vincent-Lancrin, 2006). The global number of researchers has notably increased in recent years, and this has allowed the number of researchers per million inhabitants to increase from 926 (in 2002) to 1411 (in 2015). What is interesting about this growth is that it has happened globally and not only in some central countries, and it was especially effective in countries such as China, Brazil, South Korea, and Turkey.

Infrastructures

The development of research infrastructures is one of the most relevant factors when creating and producing scientific knowledge. Infrastructure, globally speaking, is defined as the set of engineering structures and installations (usually with long durability) that are the basis on which the provision of services considered necessary for the development of productive, political, social, and personal purposes takes place[9] (Rozas and Sánchez, 2005).

The importance of infrastructures is explained by the empirical confirmation of the direct and highly significant relationship between the development of the infrastructure sector and economic growth, which is built upon the productivity of factors and competitiveness. This implies understanding that the development of modern research infrastructures is a crucial element for producing more and better scientific knowledge.[10]

Research infrastructures include human resources partnerships, covering heavy equipment and knowledge containing resources such as collections, archives, and databases. The most relevant research infrastructures that aid and collaborate in the creation and innovation of knowledge include institutes of technology, scientific parks, technological parks, technology business incubators, production and technology development centers, large scientific facilities, international campuses of excellence, and in-house R&D departments (Rozas and Sánchez, 2005).

The relevance and the impact of this infrastructure on the production of scientific knowledge are multiple: on the one hand, there is a consensus about countries' needs to increase and modernize their basic research infrastructure following international technological standards and efficiently satisfying scientific and economic agents' needs for infrastructure services (Rozas and Sánchez, 2005). The signs of progress in all areas are increasingly dependent on the capacity of producing and innovating knowledge which ultimately depends on the access to research infrastructure of the highest quality. On the other hand, excellent research infrastructure is essential to carry out

---

[9] A traditional classification of infrastructure and its connected services allows us to divide infrastructures in four large groups, defined by their objectives: (i) *economic development* (transport, energy, telecommunications); (ii) *social development* (dams and irrigation canals, drinkable water systems and plumbing, education, and health); (iii) *environmental protection, recreation, and leisure*; and (iv) *access to information and knowledge* (Perotti and Sánchez, 2011).

[10] The most important research facilities at the international level in 2019 were Los Alamos National Laboratory (US), CERN (France), Lawrence Berkeley National Laboratory (US), Bell Labs – Nokia (US/Finland), French Alternative Energies and Atomic Energy Commission (France), Chinese Academy of Sciences (China), LIGO (US); Research at Google (US), Fraunhofer Society (Germany), Broad Institute – MIT (US) (Jalan and Kopchia, 2020).

cross-border research and innovation, which is increasingly necessary for facing large global challenges. Lastly, research infrastructure also plays a crucial role in the training of young scientists and engineers and in the attraction of scientists and students from universities, research institutes, and industry from all around the world, as they offer stimulating research environments for researchers from many countries and fields.

Currently, many international actors are carrying out ambitious research infrastructure investment plans and policies. In this sense, one of the most interesting cases is the European Union, which has created a space of reflection on European policies related to scientific infrastructure, called *European Strategy Forum on Research Infrastructures* (ESFRI).[11] The ESFRI is seen by the European Union as a strategic tool to develop Europe's scientific integration and to strengthen its international outreach: "The competitive and open access to high-quality Research Infrastructures supports and benchmarks the quality of the activities of European scientists, and attracts the best researchers from around the world" (ESFRI, 2019).

Publications and patents

Lastly, scientific publications and patents must also be highlighted as new factors of scientific knowledge production. Unlike the previous production factors (i.e., R&D, researchers, and infrastructures), the inputs of scientific research, scientific publications, and patents are considered the outputs of the process. However, these two factors have become very important indicators when analyzing the generation of science because, in multicausal relations, aside from being output, they can also be considered as an input for the creation of new scientific knowledge.[12]

Scientific publications represent a key moment in the production of scientific knowledge, as they make the results of research official and publicly known. Thanks to publications, the knowledge confined to a laboratory or a university is validated by other members of the scientific community, and it penetrates

---

[11] The ESFRI, a product of the Lisbon Strategy, was created in 2002 to provide support for the establishment of a coherent framework of European policies on research infrastructures and to act as an incubator and catalyst for initiatives linked to scientific installations.

[12] The Frascati Manual points out the need of incorporating the outputs of the production of knowledge as a substantial element of analysis: "the interest in R&D is more increasingly dependent on new knowledge and innovations, and on the economic and social effects that are derived from them, than on the activity itself. It is evident that indicators on the results of R&D are necessary to complete the statistics on inputs but, sadly, it is much harder to define these indicators" (OECD, 2003).

into the field of public discussion to be the object of exams, discussions, and inputs of new processes of knowledge production. When guaranteeing the transmission and accreditation of research results, the publication is an integral part of the process of knowledge creation (UNESCO, 2005).

The strong investment in R&D has allowed the growth of global scientific knowledge production, which has been reflected in the growth of scientific publications. In total, publications have increased from 1,755,850 in 2008 to 2,555,959 in 2018 (NSF, 2018). The regional distribution also shows generalized growth throughout the international system for this same period: Asia (71.7%), Africa (60%), Latin America (30%), European Union (14%), and North America (11.3%) (UNESCO, 2015) Among the countries, we highlight China's extraordinary growth (151% between 2008 and 2014), which now represents one-fifth of the global total.

For their part, patents are also used as a trustworthy indicator of the international actors' capacity to produce new knowledge that is registered as it is valuable for future applications. The growth in the number of patents in recent decades has been extraordinary. According to the World Intellectual Property Organization (WIPO), it took eighteen years (1978–96) to reach 250,000 patent applications, but that figure doubled in just the next four years (1996–2000). Around 265,800 Patent Cooperation Treaty (PCT) international applications were filed in 2019, up 5.2% in 2018 and kept its upward trend since 2010. For the first time, applicants from China filed the most PCT applications. The United States, Japan, Germany, and the Republic of Korea completed the list of the top five origins (WIPO, 2020).

Nowadays, all these production factors are an excellent resource to verify a country's generation of scientific knowledge and, at the same time, have become elements of individual or group scientific and technological prestige and also patterns that can be used to predict the development potential of a country, region, company, or institution.

5.2 Intermediation

The second distinguishable process in the field of ISR is *intermediation*, understood as a transport mechanism for scientific knowledge, from its origin to other actors, fields, and/or places for its application and final use. This process includes specific linkages between the scientific and technological sector and many institutional spheres, but it also involves more generic mechanisms for the circulation and diffusion of scientific knowledge in society.

As it happens with the production process, intermediation is also acquiring a growing scope, mainly because STI has become the main driver of a country's productivity, the competitiveness of enterprises, the welfare of citizens, and the development of people, making the supply channels of this new production of knowledge crucial.

In recent decades, scientific knowledge intermediation has evolved and changed substantially due to the emergence of new production methods and forms, to the faster production rhythm, the plurality and diversity of created knowledge, the complexity of new knowledge, the new ways of organizing knowledge, and the appearance of new tools for its transport and dissemination. As a consequence of these new trends, in the current international context, the intermediation of scientific knowledge is produced in many different ways, depending on the type of knowledge being mediated and the mechanisms established for doing so.

For this, it is necessary to carry out a wider and more integral approach to the process of intermediation, trying to avoid reductionism and, at the same time, allowing a holistic understanding of the new shape assumed by the processes by which scientific knowledge is transmitted.[13] Generally speaking, it is possible to identify at least four different ways in which scientific knowledge is being currently transported and/or mediated from its source to its final destination.

Transmission through formal education

Formal education can be understood as the most traditional way of extending scientific knowledge throughout society itself. Ever since the transmission of knowledge was institutionalized through this social mechanism (and thus stopped being a specific task of the family), it has rapidly evolved and changed as a consequence of the massification of primary and secondary education and the later expansion of higher education. The relevance acquired by scientific knowledge on an international level and the new demands and pressures suffered by education have accelerated the changes in the mechanisms by which formal education transmits knowledge. Among these new realities, we highlight the following:

---

[13] An important element to consider when analyzing the transmission of scientific knowledge is the difference between *tacit* and *explicit* or *codified knowledge*, based on the work of the philosopher Michael Polanyi (1969). *Tacit knowledge* is that which a person has incorporated without having it permanently accessible to the consciousness but making use of it when circumstances require it; it is subjective and based on experience, and given its characteristics, it is highly personal and difficult to transfer or communicate. Its counterpart is *explicit* or *codified knowledge*, which is objective and rational knowledge that can be expressed in words, phrases, numbers, or formulae independently of context.

Higher education

In the current international context, the function of higher education as a transmitter of knowledge has become more heterogeneous and complex and encompasses undergraduate and postgraduate education and research. In the midst of deep global changes and emerging tensions, higher education is now pressured to carry out a double mission:

(i) To intensify its role as a trainer of qualified personnel, a task that higher education has done historically but is now becoming more important. The most important function of higher education in general, and of the university in particular, is to produce well-trained workers who can act as vehicles for the transfer of scientific and technical knowledge to the economic and social sectors (Montuschi, 2001). The reproduction work of qualified personnel becomes one of the main activities for the new *knowledge society*, as these knowledge workers will carry out tasks linked with the creation of knowledge through basic research, both in the public and private sectors; they will be tasked with the transmission and reproduction of tacit knowledge through teaching and will also manage the transference of knowledge to the productive sector to transform discoveries into innovations.

(ii) To encourage the transference of its research work to the productive sector, a task which is often known as the university's *third mission*. In sum, the goal is understanding the transference of knowledge between the university and the company as a process where the first one is the source of basic research generation, while the second one represents the use of the invention to improve economic competitiveness and to act as a driver of social development. Structures were built to facilitate this transfer, so the interactions are more effective; for instance, the recently created *Technology Transfer Offices*.[14]

Scientific research

A second element to be analyzed is the transmission of knowledge through scientific research. Although research is grouped in the frame of activities carried out by higher education institutions, their analytical treatment must be differentiated due to two main reasons: first, because the transmission of scientific knowledge it carries out is completely different to the one done through the process of teaching and learning, and second, because public and

---

[14] Technology Transfer Offices are understood as the units charged with managing the relation between public research and society in the field of R&D, valuing the capacities and resources of public research, and acting as the technological mediator with companies and social agents.

private research centers are increasingly unlinking themselves from university teaching institutions.

The usual way of disseminating the new knowledge produced in the field of research has always been scientific publications. These are a crucial factor because they make it possible to comply with one of the intrinsic characteristics of the activity of researchers, which is the transmission of new knowledge to the scientific community and to the whole of society. At the same time, through publications, researchers make their work known and allow other scientists to use their findings as inputs for their own research.

In recent years, the changes that have taken place in the international system due to the scientific and technological revolution, plus the advances in ICTs, and especially the use of the Internet, have had strong repercussions in the field of scientific publications. This has led to important changes in the way we transfer scientific knowledge.

One of the most relevant changes is the possibility of hosting *online scientific journals*, which has caused two very significant impacts in the process of academic communication: first, the decrease in the production costs of published content and the possibility of large institutions offering online services and access to a large number of research projects; second, the great increase in the ability to search for and read studies published online, initially through specialized search engines and later through more general tools such as Google Search (The Royal Society, 2011).

A second relevant point for the transmission of scientific research is linked with the appearance of scientific journal rankings, which have a strong impact on researchers' work, in the categorization of published research and the hierarchical classification of transmitted scientific research. The largest exponent of this new of categorizing is the company ISI Thomson, which has a vast database of evaluated scientific projects and which establishes an index of the author's, the article's, or the journal's impact. These classifications determine the greater or lesser quality of scientific works published on the basis of a methodology which, although quite controversial, has been imposed in practice and which ends up prioritizing the transmission of scientific publications (Echeverría, 2008).

Lifelong learning

A new way of transmitting scientific knowledge is related to the new reality of lifelong learning. The new international context means being part of a community where the production, exchange, and transfer of knowledge are central to all facets of life. At the same time, the speed at which these changes happen causes established knowledge to rapidly become obsolete and new knowledge to appear on a daily basis, which makes a process of continuous learning necessary.

It seems clear that, shortly, whether for everyday or professional life, the demand for knowledge will be linked to the permanent need to recycle knowledge, which will mean constant updating. Technical training in a specific field can rapidly grow obsolete with the daily changes and progress, which will inevitably force us to think about new training methods. Essentially, it is a paradigm change, as education and learning will not be limited to a specific time and space but will have to continue all lifelong. This new paradigm means extending training and knowledge transmission in a person's different life cycles.

The massification of the different educational levels, together with the need for lifelong learning, is having major consequences in responding to this new demand. The most significant changes that can already be observed are the following:

− The multiplication of learning and training centers.
− The more frequent creation of nonuniversity training centers.
− The transformation of education into a continuous process, not limited to any single time and space.
− The valuing of informal learning, whose potential is strengthened by the access offered by new technologies.

Cooperation networks

Another mechanism for scientific knowledge transmission has been strengthened through the promotion of cooperation networks in the past few years, which have flourished as a phenomenon with an important impact on research and universities. These networks are mainly composed of professors, researchers, technologists, or managers that work individually or through research groups and institutions, R&D centers, companies, and any other types of organizations whose goal is to transfer scientific knowledge (Sebastián, 2000, 2010).

Following Sebastián, we can distinguish at least four types of *networks* that work as new knowledge transmission mechanisms in the international system:

− *Information and communication networks*: They are based on electronic networks used to exchange and update information on R&D topics and, although they are usually implicit in other networks, they can exist by themselves as a fast and easy mechanism for knowledge transmission. These networks

normally connect researchers from the same discipline through the use of the new digital media.
- *Academic networks*: Mainly found in higher education, they are made up of universities, departments, centers, faculty, and researchers. Through formal and informal cooperation, interuniversity alliances and agreements are formed, the mobility of students and faculty is stimulated, and joint educational programs are established.
- *Innovation networks*: They are heterogeneous and composed of many actors, and their main goal is to improve the processes of transferring scientific knowledge to the economic sector. Due to the complexity of innovation, the actors that intervene in these networks all come from very different areas and want their interactions to help produce more and better innovations.
- *Research networks*: Lastly, most networks that have been established in recent years have been in the field of scientific research, where knowledge is multiplied (thus intensifying interdisciplinary links) and diversified (creating new cross-disciplinary communities). This network organization system is characterized by its self-management, its decentralization, and the deterritorialization of their activities, which takes place in international symposia and specialized scientific journals. This is how scientific societies lose their national aspect and are diluted into international organizations. In this context, the phenomenon of *collaboratories*[15] has had a great surge in recent years due to the extensive experience that research groups have in international cooperation.

E-learning

Finally, a new way of transmitting scientific knowledge has emerged as a result of the scientific and technological advances in digital ICTs and in the way they are being applied to formal education. *E-learning* is changing the traditional teaching and learning process and promises to completely revolutionize education. Strictly, the concept of e-learning is understood as "the use of information and communication technologies to enhance or support learning and teaching in tertiary education" (OECD, 2007).[16]

---

[15] The concept of *collaboratories* was popularized in 2005 by Koichiro Matsuura (Director-General of UNESCO between 1999 and 2009), with the goal of describing an open meeting place for academics, researchers, students, and the general public interested in the creation of flexible and participative network learning (UNESCO, 2005).

[16] The European Commission includes the distance variable by defining it as "the use of new multimedia technologies and the Internet to improve the quality of learning

The advances in the speed of capacity of electronic devices (which allow people to share a larger quantity and diversity of content), the variety in digital devices (computers, tablets, smartphones), the new types of online communication (online platforms, video conferences, video calls, etc.), the appearance of new applications to automatically share content and information (Facebook, Twitter, etc.) among many other technological and digital advances are opening a new scenery of education through the usage and application of all these new digital resources in the teaching and learning process.

In recent years, all these digital tools have started to be used in the field of higher education tentatively and disjointedly and have opened a wide range of possibilities and modalities: from courses that use the web only as a complement to traditional face-to-face education, through mixed (semi-presential and/or hybrid) courses that have teaching methods based on different interactions on the Internet, to complete online teaching.

Although it is still hard to discern, the magnitude of this phenomenon due to the lack of trustworthy statistics, the global e-learning market is considered to have generated a revenue of 35.6 billion dollars in 2011, and more than 50 billion dollars in 2016 (Adkings, 2016). Likewise, the number of students opting for distance education modalities continues to grow: in the United States, the number of students enrolling in at least one online course grew from 1.6 million in 2001 to 4 million in 2007, and it increased again to 7.1 million in 2012 (OECD, 2012). This means that, in 2012, more than a third of American students were enrolled in at least one online course.

In this context of the e-learning boom, *Massive Open Online Courses* (MOOCs)[17] are creating great expectations. A large number of universities, many of them considered world-class, have started to offer some of their courses on online platforms, which allows students from anywhere in the world to have to access them at low or no cost. Since 2011, MOOCs have experienced exponential growth, both in the offering of courses and the interest by students. Currently, 800 universities around the world offer about 9,400 online courses. In 2007, 23 million students enrolled for the first time in a MOOC, which represents a cumulative 81 million students. *Coursera*, the most popular MOOC provider, has grown from 7 million enrolled people in 2017 to 12 million in 2015 and 30 million in 2018 (Shah, 2018).

---

by facilitating access to resources and services as well as remote exchange and collaboration."

[17] The acronym MOOC was coined in 2008 by Dave Comier and Bryan Alexander to refer to a new type of *online* (taught through the Internet), *open* (anyone with Internet access can take part), and *massive* (no limit in the matriculation) courses.

The development of MOOCs and the growth of e-learning in recent years are presented as new forms of intermediation of scientific knowledge, which promise significant changes in the way that knowledge is transmitted in the field of formal education.

Due to the COVID-19 pandemic, currently, more than 1.2 billion children in 186 countries are affected by school closures due to the pandemic. As a result, e-learning is rising and teaching is undertaken remotely and on digital platforms. The unprecedented COVID-19 pandemic has forced to mobilize the entire education system to the online format and has accelerated this trend of e-learning. "With this sudden shift away from the classroom in many parts of the globe, some are wondering whether the adoption of online learning will continue to persist postpandemic, and how such a shift would impact the worldwide education market." (Li and Lalani, 2020)

Although it has numerous and complex challenges ahead (quality control, financial viability, global connectivity, professors' role, etc.), the possibilities offered by this new modality, through the use of new technology, might be able to democratize teaching by reaching more places and more people around the world, which can represent a true 21st-century educational revolution.

Knowledge and technology transfer

Aside from the conventional scientific knowledge transmission offered by formal education, there are new emerging intermediation processes in the field of ISR, such as *knowledge and technology transfer*, mainly associated with the economic and productive sectors. Although the transmission of knowledge through education, research, and networks can be considered adequate for the training of individuals or the diffusion of scientific discoveries, in the new knowledge economy, new methods and processes of transmission from scientific institutions to the corporate sector have also started to emerge.

Originally, the concept of *transfer* was more specifically associated with the transformation of scientific discoveries into innovations that would later be protected for their commercial exploitation (patents, property rights, etc.). However, in recent years, the concept of technology transference started to evolve into the term *knowledge and technology transfer*, understood as a wider concept than the previous one, encompassing a greater variety of phenomena.

Increasing relevance of transference

The prominence acquired by knowledge and technology transfer is due to the expansion of the market for STI and to the increasing commercialization of science (not only technology). The driving force behind this phenomenon is the intensification of international competition in industry and business. In the current competitive context, the company needs to create new links with universities, government laboratories, and other companies, which act as new knowledge providers (Gibbons et al., 1994; Nowotny, Scott, and Gibbons, 2003, 2006).

This is not just any knowledge, but the knowledge that emerges from scientific research, generates technological developments, and ultimately leads to innovations with commercial value. Scientific knowledge in itself is no longer sufficient and scientists need not only to disseminate it within their own community but, above all, to transfer it to other actors, agents, and preferably companies, as they are the ones generating innovations and, consequently, economic and social development and general well-being.[18]

The main argument behind the rise of knowledge and technology transfer in the productive sector is based on the increase of links between strictly scientific research and the commercial management of those discoveries. In this context, scientists transfer new knowledge to the industrial sector and give it value to be used in the field of economic competition. As Echeverría (2008) points out, "Papers provide prestige, knowledge management, higher market quotas, the capacity to attract investment and, in the end, benefits. Innovations are tested in markets and societies, not in scientific journals."

In short, the process of knowledge and technology transfer is understood as a *value chain* with many links, where, first, scientists are charged with creating basic knowledge; second, a scientific community validates that new knowledge; and lastly, other knowledge managers and administrators are tasked with processing, manipulating, and transforming that knowledge into innovations with commercial value.

Characteristics of the process

This complex process of knowledge and technology transfer towards the productive sectors has some particular characteristics that define it as such. It is important to evaluate four main elements in the way scientific knowledge is transferred to the corporate sector:

---

[18] As Javier Echeverría points out, "academic science, in which scientific discoveries, theories and facts were important, is no longer sufficient. Experts in scientific policies accept the continued existence of basic research, as it allows for breakthrough innovations, but they prefer more fertile scientific research, which may generate technological advances and, most of all, innovations" (Echeverría, 2008).

- *Nonlinear processes*: The first element requires understanding that knowledge and technology transfer question the linearity of the innovating process. Innovation must be understood as a complex network process where different actors and organizations exchange information and knowledge in the field of cooperation relations and networks on a territorial and global level to produce innovations. These pieces of knowledge are combined, applied, and distributed in interactions and learning processes between diverse actors (companies, universities, technological centers, etc.).[19]
- *Sources of knowledge*: A second element to be considered is the main sources from which knowledge emerges. The Oslo Manual (OECD and EUROSTAT, 2006; OECD, 2018) identifies three types of knowledge and technology flows towards companies: first, "open information sources," which do not involve the buying of knowledge and technology nor interacting with the source (publications, professional associations, conferences, technological networks, and databases); second, "knowledge and technology acquisitions," which is the closest practice to the *knowledge trade*, and encompasses the effective purchase of external knowledge and technology without the active cooperation with the source (incorporating machines, teams, personnel, patents, licenses, among others); and, third, the "cooperation for innovation," which is the company's active participation in innovation projects shared with other organizations, that can be other companies or noncommercial institutions (Triple Helix model).
- *Actors and intervening interactions*: Another relevant point is the participation of numerous actors that interact through specific structures and instruments. As many other scientific processes, knowledge and technology transfer involves the interaction of multiple actors. *Universities* are the main actors responsible for the production of knowledge, through which they seek to increase the value of their research to give them a practical application and to acquire funding; *companies* seek to compete in the global market through the generation of innovations coming from their internal R&D and the joint projects they establish with research centers; lastly, the *State* does not only promote R&D&I activities but mainly establishes a development-friendly environment by making legal regulations and requirements more flexible and establishing an educational system that follows society's demands.

---

[19] Some specialists mention the necessity of generating a cultural and social organization linked to the new context of innovation, as it is understood that innovation is not only the result of deliberately combining scientific knowledge, qualified personnel, risk capital, institutional policies, and many other factors, but it also requires a sociocultural system that is prepared and collectively committed with these ambitious projects.

Results: The linkage between these different actors results in different strategic actions to promote the transfer of knowledge and technology such as the promotion of public–private cooperation (university–technological centers–company), the creation of interface structures between the public and private sectors (networks, clusters, technological centers, scientific parks, etc.) and the support to new, innovating technological companies. All these actions generate intermediary structures between the actors that fulfill the mission of facilitating the transfer of knowledge such as the Technology Transfer Offices and Innovation and Technology Centers.

Diffusion through information and communication technologies

Throughout most of history, scientific discovery was confined to a reduced social group, consisting of people who were granted access to this group thanks to advanced education or social status. However, with the advent of mass information and communication technologies in the 19th century, knowledge started to spread to the general public, allowing society to get to know, even in a simplified way, the applications of this knowledge to day-to-day life. During most of the 20th century, the press, radio, and television were the mass media responsible for disseminating new discoveries. However, at the beginning of the 21st century, this central role carried out in knowledge intermediation by conventional communication technologies is destined to change drastically. The scientific and technological revolution has brought along a new way of informing and communicating through new digital media. This opens an unknown scenario concerning how new scientific knowledge is diffused in the 21st century.

Currently, television is still the most popular media as a source of information on STI developments, with about 65% of the share; it is followed by newspapers (33%) and websites (32%); other media such as radio and books occupy 17% and 14%, respectively (European Commission, 2013). Statistics show that traditional mass media, especially television, are still the main mechanism for the diffusion of STI to society as a whole or, at least, the one preferred by the population to be informed of new knowledge. However, it is also true that the information that comes from the Internet, including not only websites but also social media and blogs, has grown substantially. In 2001, the percentage of people that used the Internet as an STI information source was only 17%, but by 2013 it had grown to 35%, and by 2020 it had increased to 63% of the global population (Internet World Stats, 2020). The relevance acquired by digital media for the diffusion of scientific knowledge doubled in only 15 years and tripled in 20 years. The speed and growth of digital media is extraordinary and foretell a very close future where they might become the preferred option for the majority of the population.

Dissemination by mass media

Television, press, and radio have been the main mass media chosen by the population to learn about STI topics from the 20th century to the current days: the press has been the first from the end of the 19th century; then the radio began booming in the first part of the 20th century; and, later, television became the quintessential mass media from the second half of the 20th century.

The usage of these mass media as channels for the dissemination of scientific knowledge developed scientific journalism, in charge of the mediation between the specificity of knowledge and the journalist way of speaking about it. Even though it is a minor specialty in the field of communication, the majority of large mass media have internal sectors to present STI subjects. Even big international news agencies such as Reuters and Associated Press are large disseminators of scientific, medical, technological, and environmental information, which they feed back to the press, radio, and television continuously.

Mass media are usually criticized for excessively simplifying and trivializing scientific information and tending to turn scientific and medical news into a spectacle. However, it is also true that the volume of scientific news that appeared in mass media has had a significant increase in recent years, giving it a much bigger and greater space and visibility in public opinion.

Development of digital media

A new intermediation mechanism has started to develop due to the changes caused by the scientific and technological revolution and the emergence of new digital information and communication technologies, which are having a strong impact on the way knowledge is disseminated.

Vladimir de Semir (2003) points out an event in global media that should be considered a milestone in the evolution of the media: "July 4th, 1997 and the following days marked a new milestone in this story. This time, the protagonist media was not radio nor television: the return to Mars through the Pathfinder probe and its spectacular mini rover was seen by 45 million people through the Internet, which made it the greatest event in the then-short history of the web." For the very first time, digital media were winning the battle against conventional media, due to the versatility and the variety of options that the Internet offers over the standardized and passive information offered by conventional television.

The development of digital media has made the advent of a double paradigm evident: on the one hand, the *immaterial* paradigm, representing the independence from hardware; on the other hand, the *network* paradigm, allowing for the possibility of diffusing knowledge on a global scale. Both are progressive and growing phenomena, which have profound implications for the form, speed, and extent to which knowledge is disseminated throughout the international system (UNESCO, 2005).

Immaterial diffusion
In the current context of the international system, the process of knowledge diffusion will be increasingly mediatized by digital information and communication technology systems. Information's transmission speed has increased and the use of computers, mobile phones, satellites, and other devices allows instantaneous access to knowledge of new scientific data and findings, the state of the environment, stock market transaction flows, sports results, the emergence of social trends and new consumer preferences. A large part of this information is stored in computerized data banks linked among themselves in interlinked networks.

Participation in this increasing flow of information not only means having better computers and fast and cheap access to terminals and connectivity but also having the capacity of solving problems through selecting data and correctly organizing it. When information is abundant, competition is not generated through the capacity of generating more but by organizing existing information in an innovative way (Gibbons et al., 1994; Nowotny, Scott, and Gibbons, 2003, 2006). In this sense, algorithms are playing a new and decisive role in the virtual world as a tool for the analysis of millions of data and for choosing the most adequate options in each context.

The extension of the Internet world has contributed to a continuous creative circulation and dissemination of information and knowledge to which no individual or institution has exclusive initiative. The variety permitted by new information and communication technologies means a strong impact on two long-standing actors such as schools and books due to the digital revolution: as the relation with knowledge changes, their monopolies are increasingly eroded (UNESCO, 2005).

Network diffusion
After knowledge regimes based on oral transmission (writing and printing), digital development as a dissemination technique has led to an unprecedented expansion of networks along two axes: (i) horizontal, accelerating transmissions, and (ii) vertical, densifying connections (UNESCO, 2005). For specialists such as Jan van Dijk and Manuel Castells,[20] the conjunction between globalization (which unifies world markets) and scientific and technical development (which

---

[20] Although the concept of *network societies* has been previously broached by other academics (Wellman, 1973; Martin, 1978; Hiltz and Turoff, 1978; Braten, 1981), it was Jan van Dijk and Manuel Castells who popularized its use through the idea of seeing our current knowledge society as a *network society*.

reduces the costs of communication and helps multiply the speed and volume of transmitted information) has resulted in the appearance of the *network society*.

The *network society* is the type of society that consciously organizes its relations in digital networks, which are gradually substituting in for the traditional face-to-face-communication-based social networks. The whole world becomes the home and workplace thanks to these digital networks, and media such as the Internet are going to be increasingly considered as the normal means of communication and diffusion by being used by a larger percentage of the population and by receiving political and economic aid (Van Dijk, 2020). In the network society, social structures and key activities will be organized around digital networks. It is not only important that such networks are organized and exist in a digital form but also that they are social networks: "The Internet already is, and will be to a greater scale, the essential means of communication and relation on which will be based a new form of society, in which we already live: what I call the network society" (Castells, 2000).

The Internet is the technological infrastructure and organizational medium that allows the development of a series of new forms of social relations that do not have their origin on the Internet but are the result of a series of historical changes that could not be developed without the existence of the Internet. Castells understands the network society as a society whose social structure is built around information networks using microelectronic information technology based on the Internet. Castells reaffirms the central role of the new digital developments in the shaping of a new network society:

> In this sense, the Internet is not only a technology; it is the means of communication that is at the base of our societies' organization, it is the equivalent of what the factory or the great corporation was in the industrial age. The Internet is the core of a new sociotechnical paradigm that is actually the material basis of our lives and our relations, work, and communication methods. (Castells, 2000, 2005)

The existence of this network society, a product of the digital revolution, has meant the development of multiple expressions and shapes that those networks can acquire, through the scientific knowledge that is diffused. Among them, we highlight epistemic research or institutional networks, digital collaboratories, the development of e-learning, online education, social media platforms, and the use of phone applications, and so on.

In sum, the Internet is a new mechanism of diffusion that has a strong impact on information access and it also has great potential to radically change the way public opinion consumes information.

Mobility and circulation of highly skilled personnel

The last process of intermediation in ISR is carried out through the mobility of highly skilled personnel, which is considered one of the most traditional and important methods of scientific knowledge transmission in the international system. Historically, and excepting certain periods (such as wars), the international mobility and circulation of students, researchers, and professors has been considered a habitual phenomenon in the global system, as it has always been one of the best methods for transporting knowledge from one place to another. One of the main contributions to the circulation of qualified personnel is the transport of *tacit knowledge*, which cannot be codified and transmitted through documentation, academic articles, conferences, or other communication channels,[21] but is best transmitted between individuals with a shared social context and physical proximity (OECD, 2008).

Movement dynamics

Specifically speaking, the mechanism of scientific knowledge transmission starts to work when the researcher, scientist, or qualified person moves from their place of origin to the place of destination. Once in a different country, people share their knowledge in their workplace, where knowledge is extended to colleagues, especially those in close contact, but it also spills over to people and organizations that are geographically close. In this sense, qualified workers act as vehicles for the transfer of scientific and technical knowledge.

The relevance of this new phenomenon was recognized by the OECD when, in 1995, it published the *Canberra Manual* (part of the Frascati Family),[22] a report whose contents were specifically focused on highly skilled personnel. These human resources are understood as essential for the development and diffusion of knowledge and are the main vehicle for scientific and technological progress, economic growth, and social development. Additionally, the mobility of postgraduate and doctoral students stands out as one of the fastest-growing

---

[21] Montuschi points out that knowledge has a high relational component and, consequently, its importance often stems from the people that possess it rather than from knowledge itself. In this sense, we can point out two circumstances in which knowledge can be specific to the person that possesses it: first, when knowledge can be qualified as *sticky*, which means it cannot be separated from the person or organization containing it and it cannot be easily encoded and transmitted; and second, when individuals play a very significant role in the mediation and transmission of the knowledge they have incorporated (Montuschi, 2001).

[22] In the past 40 years, the OECD has developed a series of documents known as the Frascati Family, which includes manuals on R&D (the Frascati Manual itself), technological innovations (the Oslo Manual), human resources (the Canberra Manual), and technological balance of payments and patents, all considered indicators of science and technology.

phenomena in recent years. Many OECD countries are benefiting from the influx of international students, as many of them decide to stay in the host country, joining the local workforce once they graduate.

Global impact

Finally, the impact that this type of mobility has on the area of ISR should be considered. The circulation of qualified human resources as a mechanism of scientific knowledge transmission has varied through the years, adopted diverse forms, and had substantial effects on both its original society and its host society. In this sense, we can distinguish three types of impact:

(i) Some experts (Pellegrino, 2001, 2004; Dumont and Lemaitre, 2005; Pellegrino, Bengochea, and Koolhaas, 2013) point out that the mobility of scientists and researchers is still done from the periphery towards the center, which is translated into a constant *brain drain* for the countries of origin. The main worries about brain drain are focused on the loss of productive work and the cost of educating workers that would later move abroad and transfer their knowledge to more developed actors and destinations (OECD, 2008). The World Science Forum (2017) has called for preserving countries' scientific capabilities against the menace represented by the strong global migrations of highly skilled personnel towards more developed countries. It is considered that this represents not only a loss for the countries of origin but also an increase in the knowledge gap.

(ii) Other specialists (Meyer, Kaplan, and Charum, 2001; Faist, 2005), however, point out that the mechanism of scientist and researcher mobility has become complex, and what we can observe nowadays is more a process of *brain circulation* that stimulates the mobility of scientific knowledge to both countries of origin and destinations. This process is a consequence of the return of qualified migrants to their countries of origin after a stay abroad, or due to a pattern of temporal and circular migration between the origin and the destination, in which qualified personnel spread the knowledge acquired in their countries of origin and establish networks, thus enabling a continuous exchange of knowledge. To best make use of this brain circulation, the country of origin must have enough absorption capacity so that, when they return, talents can integrate into the local labor workforce on a level that is appropriate for their capacities and knowledge (OECD, 2008). For this same reason, some countries (such as China and India), which appreciate the value of this *brain circulation*, assign a large part of governmental resources to attract national talent that is temporarily working abroad, with the objective of having them return to their countries of origin and start new businesses or occupy high positions in academia, government, or the private sector.

(iii) Lastly, the impact generated by *scientific diasporas* is also considered through the better intermediation of scientific knowledge from the places of destination to the places of origin. A stock of highly skilled personnel extended throughout the world can act as a channel for the transfer and the mobility of knowledge and information towards the countries of origin, even if they do not return there physically, and the social and professional links increase the chances that knowledge can continue flowing even after some individuals disappear. For that reason, it is not strange that, in some emerging economies, diaspora networks have started to play a key role in the development of STI (OECD, 2008).

5.3 Distribution

After the production and intermediation of scientific knowledge, we find a third process in the field of ISR, called *distribution*. In this sense, this process must be understood as the global distribution of scientific knowledge in different fields, places, groups, or sectors on a global scale. In parallel to the increase and diversification of the processes of production and intermediation of scientific knowledge, the distribution of knowledge also gains relevance in the international system because of the relevance, impact, and consequences that this distribution has in the global system. At this point, the study of the geopolitical and geoeconomic distribution of STI in the international system is crucial to understanding where scientific knowledge is transported to and to recognize who benefits the most and who does not from this distributive logic.

Geopolitical[23] and geoeconomic[24] distribution

Analyzing the geopolitical and geoeconomic distribution of knowledge allows us to understand the complex relationship between the production of scientific knowledge, political and economic power, and geographic spaces, and how this affects the global configuration of ISR and, ultimately, the whole international system.

Many specialized organizations and experts are interested in the distribution of scientific knowledge in the international system: Mollis (2006), quoting

---

[23] Geopolitics is defined as a method of studying foreign policy that makes it possible to understand, explain, and predict international political behavior through geographical variables (Dallanegra Pedraza, 2010).

[24] Geoeconomics is defined as the use of economic tools to achieve geopolitical goals (Luttwak, 1990).

Lander, addresses the issue by pointing out that "the process becomes more meaningful when we introduce the problem of the geopolitical repatriation of production tasks and knowledge transmission. Why and for whom is the knowledge we create and reproduce?" UNESCO (2005) also wonders if "knowledge societies [will] be societies where knowledge is shared and accessible to all, or societies where knowledge is divided." Lastly, Juan José Brunner (2010) rhetorically asks "whether new conditions are effectively creating a world with more egalitarian information and knowledge; whether the flows of ideas and publications have become more symmetric; whether world-class institutions are truly within the reach of all countries."

To respond to these concerns, we cannot avoid studying in depth the distribution of scientific knowledge within the international system. Although it is possible to identify different social factors that affect how knowledge is distributed in society (socioeconomic status, age, sex, level of education, race/ ethnicity, etc.), for this analysis in the field of international studies we seek to establish a criterion that allows us to recognize and classify the geographical places and the main actors on a global level where scientific knowledge is located. In this sense, for this research, the most interesting distribution criteria are geopolitics and geoeconomics, as they allow, through statistical data, to know the relationship between the production and transfer of knowledge with the places where this same scientific knowledge is distributed.

To continue in this analysis, it is understood that the current distribution of STI can be studied by considering a series of dimensions and indicators that offer a general vision of the geographical places where STI is concentrated in the international system. To understand a little bit more about this distribution, we must analyze the main indicators of scientific knowledge and its geopolitical and geoeconomic distribution on an international level through different dimensions (R&D, researchers, publications, patents, and universities) and according to different levels of analysis in the international system (state, regional, intraregional, or economic development).

R&D investment[25]

R&D investment is one of the main indicators that show the effort carried out by international actors to develop STI. Comparing it allows us to understand how the main investments in R&D are distributed geographically and by State actors.

---

[25] All the data presented here has been compiled by the author from statistics provided by different international organizations (UNESCO, OECD, World Bank, NSF, etc.) through their online platforms or published reports.

State distribution: The United States continues to be the global leader in R&D investments, representing 25% of the global total investments; China is currently in the second place with 23% of the global share owing to its extraordinary growth in the last 15 years. Other players that stand out for their participation in the world's total investment in R&D are Japan (8%), Germany (6%), South Korea (4%), and India (3%) (NSF, 2018). By 2013, only three States/regions (China, the United States, and the European Union, known as the *Big Three*) jointly represented 67% of the global R&D expenditure. If we add Japan and Russia to this group (*Big Five*), these five countries represent almost 80% of global R&D, even though they represent only 35% of the world's total population.

Regional distribution: Continentally or regionally speaking, R&D expenditure is essentially concentrated in three geographical areas: Southeast Asia, North America, and Europe. By 2017, East and Southeast Asia represented 42% of total R&D, followed by North America with 27% and Europe with 21%. Asia's growth stands out over the rest of the regions, making it the region with most R&D investments since 2009. The regional differences appreciated in R&D investment according to GDP are also remarkable: North America still being at the head (2.7%); Oceania with its growth (2.1%); Europe and Asia (1.7% and 1.6%, respectively); and Latin America (0.7%), Africa (0.4%), and Asia's least developed countries (0.4%) being at the lower end (NSF, 2018).

Intraregional distribution: The distribution of R&D expenditures inside a region or a continent also shows great inequalities. In the Western Hemisphere, the difference between North America (the United States and Canada) and Latin America is still very large: North America's R&D expenditure is almost ten times that of Latin America's (427,000 million/50,100 million PPP$), which means that North American countries' investments represent 92% of the hemisphere's total; the proportion of the GDP spent on R&D is also very unequal (2.7%/0.7%); lastly, data also identifies a marked intraregional difference inside Latin America, where three countries (Brazil, Argentina, and Mexico) represent 80% of the region's R&D, with Brazil representing 60% of the subcontinent's total.

There is also an unequal distribution in Europe, and it is twofold: on the one hand, the European Union represents 84% of the continent's total R&D expenditure (282,000 million/335,700 million PPP$); on the other hand, inside the European Union there are also very remarkable contrasts, not just comparing the Union's richer and poorer countries, but also between relevant countries, for example, Germany and the United Kingdom in terms of R&D investment (81,700 million/36,800 million PPP$) or the percentage of GDP spent on R&D (2.58%/1.6%).

In Africa, there are differences between one country, South Africa, and the rest of the continent: South Africa represents almost 25% of the continent's total R&D expenditure (4,200 million/19,900 million PPPS$); similarly, the percentage of GDP spent on R&D is double the average percentage of Africa and triple the average of some North African Arab States.

Lastly, in Asia: Japan and China represent more than two-thirds of the region's R&D investment (431,500 million/622,900 million PPP$); the difference in the percentage of the GDP spent on R&D is also disproportionate: Israel, South Korea, and Japan are not only the most committed countries to this spending in Asia but also in the world (4.2%/4%/3.5%), while other Asian nations, mainly the Arab States, show an insignificant interest (0.1%).

Economic development distribution: The stratification based on the level of economic development is very significantly: First, the so-called developed countries represent almost three-fourths of the global R&D expenditure and have an average investment of 2.3% of their GDP; second, we have the *emerging economies* (India, Brazil, Turkey, South Africa, and South Korea) that jointly represent 26% of the world R&D expenditure and invest close to 1.5% of their GDP in STI; last, a large part of the 50 poorest African, Asian, and Latin American countries represent only 0.3% of the world's R&D expenditure. The 10 largest R&D-funding countries of 2018 accounted for $1.789 trillion in R&D expenditures, about 84.7% of the global total, and the top 20 R&D-funding countries accounted for $1.995 trillion, 94.7% of the global total (CRS, 2020).

The differences in the distribution of STI based on the economy are overwhelming. Developed and developing countries together represent 99.7% of the world's total R&D expenditure, while the 50 poorest countries represent only 0.3%. The same inequality is confirmed by the R&D over the total GDP indicator, where developed countries often spend at least 2% if not 4%, while the least developed countries have indicators around 0.2%.

Researchers

Analyzing how researchers are geographically distributed in the international system is also of great interest, considering they are the main agents responsible for knowledge production.

State distribution: Currently, the European Union is the international actor with the most researchers in the world, with absolute numbers reaching 1,726,300 in 2013, which means the EU has the largest concentration of researchers in the world (about 20%). China has become the nation with the most researchers in absolute terms, with 1,318,000 in 2013 (19% of the world total), first surpassing the United States in 2011. For their part, the United States is the second country in terms of researchers, with 1,265,100 in 2013 (16.7% of the world total). The 54 African countries jointly hold 2.4% of the total researchers in the world, and the 20 Latin American countries represent 3.6%. China, the United States, and the European Union (Big Three) jointly represent almost 60% of the world's researchers, and together with Japan and Russia (Big Five), they hold 72%.

Regional distribution: Asia continues to increase its hold on world researchers. Its global proportion was around 35% in 2002 and has grown to 43% of the world total in 2013. This increase is a consequence of China's incredible growth (19% of the world total in 2013) and the United States' and Europe's decrease (from 23% to 16% and from 32% to 30%, respectively). Africa and Oceania still represent a low proportion (around 2% each). Latin America, as a subcontinent, represents only 3.6% of all world researchers. These differences still hold if the indicator analyzed is the R&D expenditure per researcher, like North America (278,100 PPP$) and the European Union (163,400 PPP$) are still over the rest of the regions.

Intraregional distribution: Intraregional differences are also considerable: in America, the difference between North and Latin America still shows a large gap, as North America holds 85% of the continent's researchers, leaving more than 30 Latin American countries with the remaining 15%, while the R&D expenditure per researcher is almost double (298,000/179,000 PPP$). In Europe also, the contrasts are strong, as the European Union holds more than 70% of Europe's total researchers and spend almost four times as much per researcher than the other countries in the region (163,400/54,900PPP$). In Africa, South Africa holds approximately 25% (21,400) of researchers in the Sub-Saharan Africa.

Economic development distribution: In the distribution of the number of researchers based on economic development, we also find significant differences: *developed countries* represent 64.4% of the total number of researchers and spend an average of 205,000 PPP$ per researcher (2013); *developing countries* represent 34.4% of the world researcher total and spend an average of 150,500 PPP$, and the *least developed countries* have 1.3% of the total researchers and spend an average of 37,600 PPP$.

Geography of scientific publications

Scientific publications are widely recognized internationally due to the long-standing academic tradition of transmitting and presenting new knowledge to the scientific community as well as to the general public, through this mechanism. For this reason, knowing the geographic distribution of scientific publications is a good indicator to know confidently in which countries, regions, and continents new scientific knowledge is concentrated.

State distribution: In the distribution by scientific publication production per country, China has overtaken the United States for the first time in 2016, with 462,000 scientific publications (18.6% of the world total), while the United States holds the second place with 409,000 articles (17.8%). India is in third place, with 110,320 (4.8%). The European Union represents 26.7% of the total publications. Among the member countries, Germany (4.5%) and the United Kingdom (4.3%) stand out. All these actors jointly represent two-thirds of the total of world publications (NSF, 2018).

Regional distribution: The distribution of scientific publications by region first shows Asia overcoming Europe in 2014. Asia reaches, as a whole, 39.5%, while Europe represents 39.3% of the world's total publications. America, in the third place, represents 33% of the total. Far behind are Oceania and Africa, with 4.2% and 2.6%, respectively (UNESCO, 2015).

Intraregional distribution: Important differences can also be seen on an intraregional level: North American publications represent 87% of the total continent's output. Similar numbers are seen in Europe, where countries in the European Union are responsible for 86% of the continent's publications. In Asia, publications are concentrated in East Asia (79%), where Japan and China have hegemony. For its part, Africa, aside from its low performance with regards to scientific publications, also shows an unequal distribution, with one country, South Africa, holding 30% of the continent's total output.

Economic Development distribution: Lastly, differences are also notable if we analyze the distribution of publications according to economic development: *developed countries* produced 71.5% of 2014's total scientific publications (908,960 publications), while *emerging countries* produced 32% (315,742) and the *least developed countries* produced only 0.4% of the world total (3,766).

Patents

After publications, patents are the second most recognized indicator related to scientific knowledge outputs. Due to the, normally economic, but also social and political relevance of patented scientific knowledge, it is extremely useful to be able to analyze how it is geographically distributed in the international system.

State distribution: Although in 2016, PCT applicants came from 125 countries, the majority of these applications were concentrated in a few countries. In total, applicants from China, Japan, and the United States represented more than three-fifths of the total PCT applications (62%). If we add applications from Germany and South Korea to this group, these five countries represent 76.8% of the total PCT applications. These countries' joint quota increased from 66.3% in 2002 to 76.8% in 2016, mainly due to the increase in the number of Chinese and Japanese applications. China, in particular, has registered two-digit annual growth indexes in the number of applications presented per year since 2002. For the first time, applicants from China filed the most PCT applications in 2019. The United States, Japan, Germany, and the Republic of Korea completed the list of the top five origins (WIPO, 2020).

Regional distribution: A regional analysis shows similar numbers with regards to the disproportionate distribution of patents on an international level: Asia represents 47.4% of the total patents, with North America representing 25.3% and Europe 25.6% in 2016. They add up to 98% of the world total and, as a counterpart, Africa (0.2%), Latin America (0.6%), and Oceania (0.9%) do not reach 2% together. We highlight Asia's incredible growth, going from 18% in 2002 to 47.4% in 2016, mainly due to the increase in applications from China, Japan, and South Korea, at the expense of the United States and the European Union.

Intraregional distribution: The disproportion in the State and regional distribution is further increased when analyzing the intraregional distribution: in the Americas, North America represents 99.5% of the total triadic patents, while Latin America as a whole represents only 0.5%. In Europe, there is a similar disproportion between the European Union (94%) and the rest of the continent (6%). In Asia, Japan represents 87% of the region's total. In Africa, South Africa monopolizes patented knowledge distribution with 70%.

Economic Development distribution: When studying patent distribution according to economic development, the differences are confirmed: among the 20 countries with the most patents, 18 can be included in the most developed countries, and only two are middle-income countries (China and India). In 2016, three countries (the United States, Japan, and China) jointly represented 62% of the world's triadic patents. If we include Germany, South Korea, and Russia in this group, these six countries together represent 77% of the world's triadic patents, despite being only 30% of the world population. On the other hand, the world's 50 poorest countries represent only 0.05% of the total triadic patents.

Academic institutions and postgraduate students

Lastly, it is important to analyze other dimensions that allow us to complete the study of the distribution of scientific knowledge in the field of ISR. On the one hand, it is important to know the geographical location of the most relevant universities in the international system, as these institutions are often charged with the production of knowledge and innovation, and work as an attractive element for the brightest students, professors, and researchers. On the other hand, it is also important to know the main destinations of international students, considering that the number of students that go abroad to study and decide to stay in their destination country once graduated is increasing.

World-class universities

Currently, universities are distributed and concentrated in a small number of institutions on a global scale called *world-class universities*, which are the ones occupying the first places in the rankings and have the best professors, researchers, and students.

According to the ranking by the Shanghai Jiao Tong University (SJTU), the best 50 universities in the world are geographically distributed in the following way: the United States (29), the United Kingdom (7), Japan (2), Canada (2), Germany (2), Switzerland (1), Denmark (1), Australia (1), Sweden (1), Netherlands (1), and China (1). It can be considered that the majority of world-class universities come from a limited number of countries, mostly Western. Tokyo University is the only university outside the United States and Western Europe in the first 25 spots of the SJTU ranking. If we consider that there are only between 30 and 50 world-class universities, all come from a small group of eight North American and Western European countries, with Japan, China, and Australia as the only exceptions (Shanghai Ranking Consultancy, 2020).

According to the classification by *The Times Higher Education*, the 50 best universities are spread as follows: the United States (25), the United Kingdom (7), China (4), Canada (3), Germany (3), Australia (2), Switzerland (2), Singapore (1), Sweden (1), Belgium (1), and Japan (1). This ranking also indicates that the clear dominion of the United States and Western Europe (especially the United Kingdom) and the appearance of new universities in East and Southeast Asia (China, Japan, and Singapore).

Both rankings show a geographical distribution in just ten countries, with a special concentration in the United States (between 50% and 60%). The Anglo-Western university block (the United States, the United Kingdom, Australia, Canada, and New Zealand) and the rest of Western Europe cover 93% of the first 100 spots in the world rankings. At the same time, they capture 8 out of every 10 international higher education students and have 99% of the first 100 universities and 90% of the first 500 (Shanghai Ranking Consultancy, 2020).

International students

It is also important to know the geographical distribution of postgraduate students on a global scale as their movements are increasingly relevant in numerical terms and according to the demographic and economic impact they have. Statistical data shows that, between 1997 and 2013, the number of international students more than quadrupled, going from 800 thousand to more than 4 million. In the 1960s, American and, to a lesser extent, European universities already had a large number of foreign university students. This trend was intensified in the last three decades, and currently, it is estimated that more than four million students are enrolled in foreign universities, of which approximately 80% study in OECD countries. The phenomenon is still growing, and the number of international students is expected to grow to 8 million by 2025 (UNESCO, 2015).

Generally speaking, students move from *developing* to *developed countries*, and from Europe and Asia to the United States. The majority of undergraduate and postgraduate universities choose the United States (19% of the world total), the United Kingdom (10%), Australia (6%), France (5%), Russia (5%), and Germany (5%) as their favorite destinations (OECD, 2016). Despite this, in recent years, new countries have emerged as international student destinations, mostly neighboring countries, such as Australia, China, and South Korea in the Asia-Pacific region; South Africa in Africa and Brazil; and Chile in Latin America (NSF, 2018).

Although there is proof of increasing competition between countries for student recruitment, until now, the United States still occupies a predominant position in foreign student attraction, mainly in higher education: American universities attract students from all around the world and recruit approximately a third of all foreign students, with Asians being the first main group and Europeans the second one; American preeminence is most strongly asserted in the segment of the highest qualification and relevance to national innovation capabilities; and among foreign doctoral students, the United States has an amount equivalent to the sum of all the other OECD countries (Luchilo, 2006). Additionally, the United States is also the main destination for postdoctoral students.

From this, we can conclude that the market of highly skilled personnel has a global offer, but a demand mostly concentrated in a single country (the United States), or, in the best of cases, a small group of countries (the United States, the United Kingdom, France, Canada, and Germany).

5.4 Application

A fourth operating process observed in the field of ISR is linked with the application of scientific knowledge in specific fields. In this case, the goal is to understand the final destination of the knowledge, which is produced, transported, and distributed all throughout the international system. Scientific and technological application is a key factor to determine the areas where scientific knowledge is being used and to see in what fields (economic, social, political, military, cultural, education, etc.) new pieces of knowledge are being applied. To identify these destinations, it will be necessary to analyze the following dimensions and indicators that will allow us to know in detail how the application process develops: (i) type of research, (ii) scientific fields, (iii) publications and patents, (iv) socioeconomic impact, and (v) application in the private sector.

Type of research

The first way to see where scientific knowledge is applied is through the type of research carried out. The classification most used to characterize R&D expenditure differentiates the production of scientific knowledge according to the type of research in basic, applied, or experimental development (OECD, 2013; NSF, 2018). While this categorization is questioned because the process of producing scientific knowledge should not necessarily be considered linearly in these three phases, it is still a fairly good way of understanding the motivations, expectations, and results expected from investments made in R&D:

- *Basic research*[26] is usually carried out by scientists and researchers (usually, in the context of higher education) with enough freedom to choose their own goals and whose results are not often sold but published in scientific journals. The race for achieving *applied* and *commercially valuable* knowledge has reduced the investment in basic research in many countries (the United States, Canada, or Russia), but it still holds in places such as the European Union. The proportion in R&D expenditure aimed at basic research oscillates between 5% and 25% (France 24%, the United States 17%, Russia 17%, and China 5%) (NSF, 2018).
- *Applied research*[27] develops ideas and transforms them into something operational, as the results of these types of research are focused on a single product or a limited number of products, methods, or systems that are often patented. The percentage of R&D expenditure destined for this

---

[26] Defined as "experimental or theoretical work undertaken primarily to acquire new knowledge about observable phenomena and facts, not directed toward any particular use" (OECD, 2003).

[27] Defined as "original investigation to acquire new knowledge directed primarily towards a specific practical aim or objective" (OECD, 2003).

type of research varies from 11% (China) to 43% (the United Kingdom). The United States spends about 20%, but in absolute terms, the volume of North American investments is well above any other global actor (NSF, 2018).

– Finally, *experimental development*[28] consists of projects that make use of existing knowledge for the production of new materials, products, processes, or services, or the substantial improvement of those already existing. This type of research is the one that has grown the most and where countries have shown the most interest when compared with the other two. The average investment in experimental development is between 60% and 70%, and the majority of great world powers spend most of their budget in this field. China exhibits the highest percentage with 84%, followed by the United States and Japan with 64%, and South Korea with 62%. In absolute terms, China and the United States are the largest investors.

Scientific fields[29]

A second dimension to know how scientific knowledge is applied in the international system is by its distinction by scientific areas and fields, which allows us to describe its application in the different areas of science. This indicator is especially interesting, as it is useful for understanding in which specific scientific fields the major States are applying their investments in STI:

– Among scientific fields, the main beneficiary of the budget investment is the field of *engineering and technology*: on average, it receives more than 50% of the total investment, and in some particular cases, it can reach three-fourths of the total. In the most significant cases, we find China (80.7%), Russia (70.5%), and South Korea (69.8%) in 2010.
– The second field of interest is *natural sciences*, where the United States spends approximately 20% of its R&D investment. The most relevant examples are Australia (30.5%) and Russia (20.3%) in 2008.
– The third field is *medical sciences and health*, where the average investment of 10% of total R&D is stable. The State actors that spend the most in

---

[28] Defined as "systematic effort, based on existing knowledge from research or practical experience, directed toward creating novel or improved materials, products, devices, processes, systems, or services" (OECD, 2003).
[29] It should be noted that that many countries do not facilitate national data on the R&D expenditure per scientific field to the OECD, as is the case with the United States, the United Kingdom, or Germany, which makes international comparison more difficult.

this field are South Africa and Australia (14.9% and 14%, respectively, in 2008), Spain (13% in 2010), and South Korea (10.8% in 2010).
- The investment in the field of *agricultural sciences* stands somewhere between 2% and 14%, although the historical evolution shows a strong trend towards the former. Countries with large investments linked to agricultural production are, for example, Argentina (14.1% in 2009), Australia (9.3% in 2010), or South Africa (8.2 in 2005).
- *Social sciences and humanities* are the areas with lesser investment when compared with other fields. Countries that stand out are Argentina (5.7% in 2009), South Africa (2.8% in 2008), Australia (1.5% in 2009), South Korea (1.2% in 2010), and Russia (1.3% in 2010).

In the current international system, States decide through their investments which are the most strategic fields in which research can be carried out, leading countries to specialize in specific fields: South Korea, China, and India dominate *engineering and chemistry*; the United States maintains its hegemony over *medical sciences and health*; France and the United Kingdom are leaders in *social sciences*; Brazil focuses on *agricultural sciences*, while Russia is interested in *physics, mathematics, and geosciences* (UNESCO, 2015).

Publications and patents

A third dimension is represented by the analysis of publications and patents based on scientific areas, which allows us to understand the result of research according to the specific areas where they have been applied.

On a global scale, the fields with the largest number of scientific publications are *biological, medical, and other life sciences* (representing 38.6% of the world total), followed by publications in the field of *engineering* (18.4%), and then those linked to *physics* (8.7%), *computer technology* (8.2%), *chemistry* (7.9%), and *social sciences* (5.3%).

A country-based analysis confirmed the growing specialization: approximately 50% of the publications in the United States are linked to the fields of *biological sciences, medical sciences*, and other *life sciences*, and a significant number of publications also appear in the field of *psychology and social sciences* (NSF, 2018); the European Union also focuses its publications in the field of *life sciences*, with more than 40% of the total publications; China, for its part, is focused in *engineering, chemistry, and geoscience* (40% of its total), while India specializes in *computer technology* (where it is the world leader) and *engineering* (24%) (NSF, 2018).

The applications for triadic patents show a growth in the fields of *computer technology, information technology, electrical machinery, digital communication, energy, and medical technology* (WIPO, 2020). Concerning USPTO applications, of the total number of patent applications in 2016, 37% were linked to new ICTs (*information technology*, 14%; *digital communication*, 10%; *telecommunications* 4%), followed by 16% to *chemistry and health*.

Socioeconomic impact

Another valuable dimension for a better understanding of the process of scientific knowledge application is how governments make investments depending on the socioeconomic objectives they pursue. These indicators clearly show each of the State's ultimate goals concerning the application and use of new knowledge. For this, it is relevant to highlight two categories: R&D expenditures linked to the field of *defense*[30] and *civil* expenditures linked to the rest of the socioeconomic objectives.

– Traditionally, R&D expenditure in defense has been high in most national budgets. The analysis of investments shows three significant trends: (i) a small group of countries maintain high R&D investments in the defense and military security sectors, such as the United States (the largest world investor, with 51.7% in 2017) and Russia (52% in 2003); (ii) a large group of countries has reduced their defense budgets, such as the United Kingdom (16% in 2016), France (6.4% in 2016), and Spain (1% in 2016);[31] and (iii) a new group of countries has increased their military R&D budgets, such as China, Turkey, and Saudi Arabia.

The socioeconomic-driven R&D expenditures (excepting the defense sector) allow us to identify the objective behind the main international actors' application of scientific knowledge in the civil sector. The most relevant *socioeconomic* objectives include the following:

– *Economic development*,[32] which has traditionally been one of the areas that has received the most funding, though somewhat reduced in the last few years. Currently, OECD countries invest an average of 21% of their global R&D budget in this objective (2016). Among the highest-investing

---

[30] Defined as "all defence R&D financed by government, including military nuclear and space but excluding civilian R&D financed by ministries of defence" (OECD, 2013).
[31] All the statistical data are from the national R&D budget focused in the defense sector, as a percentage of the government R&D expenditure (OECD, 2018).
[32] *Economic development* promotes "agriculture, fishing, forestry, industry, energy and infrastructure, and general land-use planning" (OECD, 2018).

countries are Australia (27.5% in 2016), Japan (26% in 2017), the United Kingdom (18% in 2016), and France (17.2% in 2016).
- *Health and environment*,[33] where the majority of countries have proportionally increased their R&D expenditures, thus becoming, in many cases, the objective with the highest investment. From the largest to the smallest, we have the United States (52.7% in 2017), the United Kingdom (33.2% in 2016), and Australia (28% in 2017). OECD countries spent an average of 23.4% of their total R&D budget in 2015.
- *Education and social programs*,[34] where investment is significantly lower than other objectives. Among the highest-investing states are Greece (14% in 2016), Denmark (8.4% in 2017), South Korea (8.2% in 2016), Norway (7.5% in 2017), and Italy (7% in 2016). Below these levels, we find the United States (2.5% in 2017), Australia (2% in 2016), and Japan (0.6% in 2017). The average investment of OECD countries (3.5% in 2016) is low in these objectives.
- Lastly, *space programs*,[35] where the highest-investing countries are the United States (17.6% in 2017) and Russia (18% in 2003), while the average of OECD countries is 8% (2017).

Private sector

Finally, considering the important role large transnational companies have acquired throughout international systems, along with the fact that in many countries, they are the main R&D investors, it is important to analyze the fields in which the private sector is investing. Contrary to States, companies have no obligations to invest according to the society they represent; the driving forces behind these investments are their economic and strategic interests, so their expenses and applications in STI are quite different.

Currently, companies' main objective is to develop scientific knowledge that allows them to innovate in marketable products, processes, and services. For this reason, companies are seeking higher levels of competitiveness by making a large part of their companies and foreign subsidiaries R&D&I investments in the field of intensive industrial research, in sectors such as pharmaceutical products and hardware, in software technology, and in new

---

[33] *Health and environment* promotes "the protection and improvement of human health, the control and care of the environment and the exploration and exploitation of the earth" (OECD, 2018).
[34] *Education and social programs* finance R&D in "education, culture, recreation, religion, communication technologies and social and political systems" (OECD, 2018).
[35] *Space programs* finance civil R&D on "topics linked to space" (OECD, 2018).

emerging technologies (robotic, biotech, artificial intelligence, etc.). Aside from new products, investments are prioritizing applications in the development of software and services complementary to their traditional offers. Among the most innovating international companies by economic sector, the following are the top 10: three in the *software and internet* sector (Alphabet, Amazon, and Microsoft), three in the *computer science and electronic* sector (Apple, Samsung, and Intel), three in the *pharmaceuticals* sector (Roche, J&J, and Merck), and one in the *automotive* sector (Volkswagen) (PwC, 2018). For their part, in the top 25 companies with the highest R&D investments by economic sector, we find eight in the field of *health and pharmaceuticals*, six in the *automotive* sector, six in the field of *software and internet*, and four in the field of *computer technology, technology, and electronics* (PwC, 2018).

Additionally, *artificial intelligence* becomes to play a significant role in business competitiveness. Fast-growing companies of the past decade were technology companies that also invested heavily in artificial intelligence. Most of them from the United States and China—Google, Apple, Facebook, Huawei, Alibaba, and Amazon (Rathenau Instituut, 2020).

In short, there is a clear and consistent trend, by the majority of the actors in ISR, of creating a process of scientific knowledge application characterized by (i) *experimental development* and *applied research*; (ii) technical disciplines such as *engineering, technology, and health sciences*; (iii) the *specialization* of the applications by countries and by disciplines; (iv) scientific publications in *biological and medical sciences and engineering*; (v) patents linked to *information technology, electric machinery, digital communication, and medical technology*; (vi) the development of *software and complementary services for consumers*; (vii) socioeconomic objectives linked with *economic development, natural sciences, health, and environment*; (viii) sustained expenditures in *defense*; (ix) a private sector more inclined to invest in R&D and often supplanting the public sector, with strong investment in the *technological and digital sector*; and (x) heavy investment in *artificial intelligence*.

5.5 Governance

The fifth and last of the processes distinguishable in the field of ISR is *governance*. This mechanism is the operative process through which many international actors carry out tasks linked to the planning, organization, and execution of public policies and private strategies for the development of STI.

Traditionally, the control over these processes within the international system was held by Nation-States, who autonomously designed public policies in science and technology through their internal organizations (ministries, directorates, and/or agencies) with limited intervention and validation from other actors, except minor participants from industry and university. Although companies had advanced in the *management* of knowledge as the main strategy in the development of competitiveness and wealth generation, these processes were carried out specifically in the private sector, while they let the State assume the leading role of the *governing* of scientific knowledge.

However, many of the changes that have taken place in the international system in recent decades have also had consequences in the planning and organization of STI, with the arrival of new actors that have directly linked themselves with the production of scientific knowledge and have started to demand higher participation in the process, which resulted in a much more diverse configuration and government and management structure. Experts such as van Kersbergen and van Waarden (2004) argue that the best way of describing these deep changes, which are fundamentally different from previous mechanisms, is through the word *governance*.

The key in this new process of policy planning and execution in STI has been precisely the transition from organizational schemes linked to the *government* and the *management* of knowledge to a new way of articulating and coordinating scientific knowledge through *governance*. Although occasionally the concept of *governance* is confused or used indistinctly with *government or management*, it is a distinct phenomenon that emerges as a response to the need of explaining new processes of governability.

As Nicolás Mariscal (2006, 2011, 2017) points out that the old term *governance* became more current and relevant and acquired new meanings at the start of the '90s, in part due to the fact that it has had many, not rigorous uses. James Rosenau (1990, 1995, 2003) understands that the concept of *governance* is a wider-encompassing term than *government*, as it includes governmental institutions, but also informal and nongovernmental mechanisms, through which people and organizations in their respective fields advance and satisfy their needs and wishes. In the same line, Francesc Morata (2004) sees governance as a "new style of government" involving public and private actors in the political process and a larger degree of cooperation than the classical State-headed hierarchic model.

In the field of ISR, the different actors, relations, and international processes are distributed throughout multiple fields and/or arenas of operation, which means the governance of STI is carried out simultaneously at different levels.

Public policies

In the current international system, the national level still has an important role in the planning and organization of STI. Although nowadays other actors are indeed taking a very active part in this task, it is also true that the State is still a sort of catalyst in the planning process, on top of which pivots the creation of knowledge through the development of government-made public policies. In this context, States still are central stakeholders and they have important responsibilities when designing, funding, executing, and monitoring these policies.

The relevance of the State for STI goes back to the '30s, which marked the beginning of a new commitment of the States in the planning and organization of scientific knowledge (*Big Science*). From this point forward, States became the main agents responsible for governing and managing the production of scientific knowledge, a strategic resource for national development. This task was carried out through the generation of public policies linked to science and technology, also called *scientific policies*, which, analogously to other public policies, expressed a government's interest in the field of science (Albornoz, 2007).

This state planning process has been modified over time, mainly because the linear vision of scientific knowledge production, which appeared at the beginning of *Big Science*, has been, for several decades, strongly criticized, rebuked, and partially replaced by much more complex knowledge generation models (in processes as well as actors and interactions), which, in part, has meant the readapting of the State to a new context where it shares responsibilities with more actors of the international system.

Whatever the case, Nation-States are still considered to be a core actor for the articulation of interests, the planning of processes, and the executions of tasks linked to STI, where it becomes a sort of mediator in the discussions generated between different actors that pursue their own interests. Currently, States, sub-State entities, companies, think tanks, epistemic communities, and universities have substantial interests in the creation of these public policies, which in turn creates a subtle game of powers and crossed interests.

Nowadays, the majority of countries carry out ambitious public projects to increase their investment in R&D, which has allowed an increase in the R&D expenditure in the majority of countries with the objective of generating favorable conditions for the development of STI. The strong investments carried out by States in the international system follow the logic of considering scientific knowledge as a key element for economic and social development.

The most scientifically developed countries have been the first in pushing strong public planning for state investment in STI: the United States established its Office of Science and Technology Policy in 1976, where it designs and executes all its STI projects; the European Union started its plans in the mid-'80s; Japan designed its first five-year plan in 1996 and is currently on its fifth; South Korea introduced its Science and Technology plans in 1996; China launched its Medium- and Long-Term Plan in Science and Technology in

2006; and India, which has a long tradition of quinquennial plans started in 1951, is now in its 13th edition.

The least economically developed countries have also started to carry out STI plans. The majority of them have become aware of the necessity of considering scientific knowledge as a strategic investment and have developed their own public funding projects. Despite the economic hardships many of these countries face, now STI investment is considered a priority national strategy:

- In Latin America, aside from Brazil, which is the region's largest R&D investor, countries such as Argentina, Colombia, Mexico, and Chile are carrying out strong investment processes, as are other smaller countries such as Panama, Costa Rica, and Bolivia.
- Central Asian countries, after the post-Soviet era, have reorganized their STI plans and have driven new R&D investments, as has happened in Estonia and the Republics of Kazakhstan and Uzbekistan.
- Arab countries, with historical deficits in their STI systems, are showing a renewed interest in increasing their R&D expenditures with very ambitious public projects and plans, such as in Qatar, the United Arab Emirates, and Saudi Arabia.
- Finally, Africa, despite economic adversities, shows a particular interest in the development of R&D investment systems as a core element of development. Three African countries (Ghana, Mali, and Senegal) are of particular interest, as they are investing over 1% of their GDP in R&D.

In recent years, the increasing relevance of new emerging technologies (artificial intelligence, robotics, biotechnology, etc.) and their direct impact on the international system are intensifying the role of public STI policies. Most developed countries have started to create long-term strategies with the goal of investing in new emerging technologies, seen as tools of economic, social, and geopolitical development. In the past six years, 8 of the 10 most developed countries have launched national programs linked with the usage of STI for the development of the *digital economy*: first was Germany, with its 2011 *Industrie 4.0* program; afterward, in 2012 and 2013, the United States, Italy, and France; later, East Asian countries such as South Korea, Japan, and China joined this trend. The Chinese strategy *Made in China 2025* aims to improve its competitiveness in ten emerging technologies applied to the new economy. India, through its *Make in India* initiative (2014), and Russia, with the *Advanced Industrial Technologies Program* (2017) are also developing long-term strategies. Finally, Australia, Canada, and Spain are following this same path by planning national strategies with analogous objectives (WEF, 2018).

The main goal of all these public policies is investing in STI as a key tool for economic and social development through measures linked to strategic scientific sectors. All these programs apply very similar policies that, in general, are characterized by (i) a strong investment in higher education, (ii) talent attraction policies, (iii) the intensification of R&D&I investment, (iv) a strong commitment to developing emerging technologies, (v) tax incentives to support the private sector, and (vi) a legal framework to regulate and protect the production of new scientific knowledge.

Local governments

In recent decades, governance on a local level has emerged as a strong and increasingly relevant space in ISR. In parallel to the rise and boost that local actors and fields have had in the last few years, scientific knowledge governance has also started to develop in sub-State entities such as cities and regions (also called provinces, states, or autonomous regions). Public policies geared towards the encouragement of STI in the context where they belong, but with a clear connection to the global system,[36] have started to appear in these novel spaces of governance.

In recent decades, the appearance of concepts such as the "global city" (Sassen, 2001) or "glocalization" (Robertson, 1995, 2015; Bauman, 1998) to describe the new relevance of local spaces connected to the international system has received special attention. Manuel Castells (2000, 2005) extended this vision to the field of science and technology when he introduced the so-called *new connectivity* perspective, which holds that, in advanced knowledge and higher education networks, the central nodes (cities and regions) play a strategic role because they control the means of production, communication, and value creation, determine the programs and protocols that govern the flows and activities of the participants and, most importantly, have the power to create networks or to connect different networks among themselves, organize their cooperation and avoid competition. This point of view has been expanded with studies that prove

---

[36] From the economic perspective, the concept of *local development* has been particularly popular to explain: "the process of the transformation of economy and local society, geared to overcome existing difficulties and challenges, to improve the living conditions of its population through a concentrated and decided action between different local socioeconomic actors, both public and private, through the encouragement of the entrepreneurial skills of the local company and the creation of an innovating environment in the territory" (Pike, Rodriguez-Pose, and Tomaney, 2007).

how scientific knowledge ends up being distributed through these same centers and nodes (García Guadilla, 2010; Innerarity, 2011, Knight, 2018; Sassen, 2017; Van Dijk, 2020).

Sub-State entities, comprised of cities, regions, and local government, have recently acquired enough power and competencies to design and execute public policies and plans in the fields of STI and education in their respective jurisdictions. This has allowed them to develop a level relatively autonomous from the State's sphere, which links local actors with the production of scientific knowledge. At the same time, the local level is one of the most favorable for involving civil society in governance because of the proximity with local actors. "The local level has a high degree of self-organization. Due to its closeness to citizens and the comprehensiveness of the problems it faces, it is the field in which civil society participation can be best increased" (Innerarity, 2011). What is relevant and innovative in this governance process is that STI planning and organization are directly debated by local actors and, as a consequence, they design policies that contain and consider the specificities of each local field. These plans have become common in the majority of cities and regions that opt for projects that involve the closest community and have a direct impact on the local sphere.

The local level has become a new arena for actors interested in STI, who see it as a place of negotiation and debate for the creation of sub-State public policies. It is a new space that stimulates the interaction through many diverse activities: (i) contact with local agents, (ii) intermediation with superior administrations (on a state, regional, or international level), (iii) connection with local cities and territories, and (iv) a more active participation in the planning of public policies that directly affect them.[37]

In the current international context, more and more cities and regions become geographical units interested in designing and controlling their own public policies in STI. These cities and regions seek to promote the exchange of knowledge between institutions and organizations grouped in their environments, they offer a larger concentration of talents and are able to encourage more knowledge-intensive economies because the region or city offers an attractive place to work, invest, and research.

---

[37] The Royal Society (2011) notes that, although processes in the sub-State city and region levels or arenas have become more relevant in governance processes, they still lack trustworthy statistical data.

Regional initiatives

Intending to achieve common objectives and benefits, many Nation-States have progressed in regional cooperation and integration processes, through the harmonization and/or unification of their economic, social, and cultural policies, thus transforming the regional idea into a new and profitable governance space. Although it is not a strictly new phenomenon in the field of international studies, in recent decades, there has been a proliferation of regionalization and regional integration processes, which have stimulated the linkage between actors located in the same region or continent.

Currently, regional processes span the globe with different levels of breadth and depth in terms of the agenda items addressed and the actors involved in their discussion. In the field of STI, the cooperative agreements reached are notable, in many cases managing to establish a true synergy and the harmonization of interests and joint action plans linked to the development of scientific research.

Governance in STI in the regional sphere has been specially promoted and spearheaded by the intense cooperation developed within the European Union, which has started a proliferating path of planning and regional organization of its science and technology policies. The European Union has been one of the international actors that has shown the most interest and concern for establishing joint strategies for the advancement of scientific knowledge. There have been several milestones in recent years:

- In the *Lisbon Strategy* of March 2000, the Heads of State of all member countries of the European Union agreed on a new strategic objective for the regional process "to make the EU the most competitive economy in the world," for which the investment in higher education and scientific research was prioritized, as they were seen as key elements for economic and social development (European Commission, 2001).
- In 2009, the *Treaty of Lisbon* was revised and ratified through the *Europa 2020* strategy, where a plan of action on the policies that the member countries of the European Union must apply to generate growth was proposed. Among these recommendations, the concept of *intelligent growth*, directly linked with R&D investment and the development of higher education and quality research, is explicitly mentioned (European Council, 2010).
- Shortly afterward, the report of the Reflection Group to the European Council (European Council, 2010) was released, suggesting some recommendations, published in the work "Project Europe 2030: Challenges and opportunities," where some goals to be met by the year 2030 are

established. Among the many recommendations, we highlight the chapter on *growth through knowledge*.

Even with nuances, the result of this complex and negotiated regional process of STI governance by the European Union has had amazing results: (i) the *Bologna Process* (1999) led to the creation of the *European Higher Education Area*; (ii) the establishment of the *European Research Area (ERA)*, a space to host the greatest European and international talents; (iii) the establishment of the *European Commission's Framework Program (FP)* as the main tool through which the European Union's member countries collectively invest in scientific knowledge; and (iv) the formation of the *ESFRI* in 2002 to support in the establishment of European modern research infrastructure policies.

The Southern Common Market (MERCOSUR), the regional bloc that incorporates many South American countries, is another good example of regional STI governance. Inspired by the European Union's model, the MERCOSUR arose as a regional cooperation process that aims at full economic, political, and social integration. It is therefore not surprising that STI agreements have been in place since its inception. The first step took place in 1992 when the member States created the *Reunión Especializada de Ciencia y Tecnología (RECYT)* with the goal of creating scientific and technological policies for each region (Zurbriggen and González Lago, 2010). This process was further deepened with the creation of the *Reunión de Ministros y Altas Autoridades de Ciencia, Tecnología e Innovación* (RMACTIM) as a political body for scientific and technological collaboration between member States. Among the strategic goals for the future governance are (i) promoting the advancement of knowledge in strategic areas, (ii) promoting mechanisms that would lead MERCOSUR countries to become a knowledge society, (iii) training of human talent and the improvement and outfitting of scientific infrastructure, (iv) promoting the creation of knowledge networks, and (v) promoting the usage of ICTs in the democratization process.

New international sphere

The international level is another relevant field of governance where different actors interact and establish relations outside their closest environments. The conceptualization of an international arena implies the existence of a space for action for STI actors that transcends their usual national frontiers. Inside this international level, we find a multitude of suprastate, intraregional, continental, trans-sovereign, transnational, or global processes (both formal and informal), all of them with their own dynamics.

In this global context, many actors in ISR act to create, manage, and transmit STI, generating multiple and diverse types of interactions. Among the main types of links established on this level are (i) bilateral and multilateral relations between Nation-States; (ii) interactions between States and IGOs; (iii) agreements between States, companies, and universities; iv) contacts and channels linking companies or universities to each other; and (v) special events where the majority of international actors participate and interact. Once the possibility of establishing links between different actors on an international level is open, contacts, modes, actions, and types of relationships seem infinite.

As a result of this multiplicity of new interactions in the international arena, it is now possible to distinguish some efforts (mostly governmental) to establish global governance in STI. In this sense, the activities organized by the United Nations, through its specialized agency for education, science, and culture (UNESCO) are especially relevant. This organization carries out multiple tasks of planning, coordination, and articulation of research and advice regarding scientific knowledge on a global level, in the certainty that it will be increasingly important as a strategic resource and that "the struggle for these cognitive resources will be a key political and economic element" (UNESCO, 2005). In parallel, other intergovernmental organizations such as the World Trade Organization (WTO), the General Agreement on Trade in Services (GATS), and the World Economic Forum position themselves as important new spaces of global governance of STI.

This increasing phenomenon of STI global governance has caused many actors, especially researchers themselves, scientific personnel, epistemic communities, and scientific diaspora, begin to promote initiatives tending to channel multiple dynamics and linkages that have been carried out on an international level towards a more planned, articulate, and organic governance. In this sense, in recent decades, one of the main landmarks has been the *World Conference on Science*, taking place for the first time in July 1999 in Budapest, under the auspices of UNESCO. The *Declaration on Science and the Use of Scientific Knowledge*, which established a set of principles linked with the proper use of science for global development, was approved and proclaimed there. As a corollary of this conference, the biennial *World Science Forum* was created, where ideas on the evolution of science on a global scale are evaluated, debated, and exchanged.

In November 2019, the ninth and latest edition of the *World Science Forum* took place in Hungary. In its final declaration, titled *Declaration of the 9th World Science Forum: Science Ethics and Responsibility*, the existence of a *new age for science* on an international level was established, with unprecedented political, economic, and social consequences; the *new role of knowledge* was mentioned in a new age of humanity's history, aiming to tackle the great global challenges; *cooperation and competitiveness* in science were promoted; and new mechanisms of cooperative governance were established to channel this process for the benefit of society as a whole (WSF, 2019).

Part 3

EXPLICATIVE FRAMEWORK

Chapter 6

EMERGING REALITIES

The systemic study carried out up to this point on ISR (Part 2: Analytical Framework) has allowed us to discover and examine the global international context (Chapter 2), to identify the main international actors taking place in its dynamic (Chapter 3), to recognize the different interactions established between them (Chapter 4) and, finally, to discover the operative processes and mechanisms that develop within them (Chapter 5).

As a result of this analysis, in this chapter, it will be possible to identify and analyze some *emerging realities*[1] that have arisen precisely as a result of the links established between the context, the actors, the relations, and the processes within ISR. These emerging phenomena become distinctive characteristics of ISR, with a great impact on the entire international system.

ISR is understood in this research as a system of interrelated actors and, as far as some of their main characteristics are mere results of the property of its members (the so-called *hereditary or resulting properties*), some derive from the relations among the actors, but without the actors possessing them themselves (the so-called *emerging or collective properties*).[2] The interactions between actors in the field of ISR produce emerging patterns of conduct that, even though no one has planned, wished for, or sought them out, proliferate as global characteristics and become trends and macrotrends that feed back to the rest of the international system.

For this reason, in this chapter, it will be important to identify, characterize, and explain which most relevant emerging realities are stemming from the study on ISR and to discern the possible consequences they might have for the new world order.

---

[1] Bunge defines emerging properties as "qualitative novelties," and understands that "a system is a complex object that has global properties and acts as a whole because its components are joined together" (Bunge, 2003, 2009).

[2] This means *emerging realities* cannot be reduced to the actors of the systems; they are properties these components would not have in isolation and that cannot be explained as elements of the system, but only as the product of the interaction between them.

6.1 New Emerging Phenomena

Through the extensive study of the systemic parameters of ISR, it has been possible to distinguish some emerging realities. Among them, we can mention the rise of *STI diplomacy*, the new struggle for *global talent*, the emergence of international *cooperation networks*, the development of *scientific diasporas*, the appearance of new *methods of production* and *forms of intermediation* of scientific knowledge, and the emergence of a renewed and intense *competitive race* between the different international actors to achieve a higher development of their STI systems.

On top of the emerging realities that have appeared in the previous analysis on ISR, it is also possible to identify other very important and significant emerging phenomena for ISR and the global international system. For this reason, these new emerging realities require a more in-depth analysis and explanation, which will allow us to discover their main characteristics and their possible consequences for the entire international system.

6.2 Knowledge Gap

Some of the most notorious emerging realities that can be observed in ISR are the so-called *knowledge gap*, which arises as a consequence of the unequal distribution of technological and digital resources that, with time, has extended to the distribution of scientific knowledge itself. This emerging phenomenon has a great impact on the way STI is distributed on a global scale and the way this affects and deepens other previously existing social gaps. While, on the one hand, it is thought that scientific and technological discoveries and applications can bring great social benefits by helping fight against humanity's great challenges, on the other hand, the concern for the unequal way STI is distributed and the consequences this will have in the future is extended.

Types of gaps

The unequal distribution of STI has given rise to the emergence of new gaps in the international system: the *digital divide*, the *cognitive divide*, and the *scientific divide*. Generally speaking, these concepts have evolved from the precursor idea of a digital divide to the more recent ideas of cognitive and scientific divide, as this process has grown in its reach, relevance, and complexity. These gaps coexist and feed back into each other, thus forming a concerning emerging reality of the acceleration and intensification of inequalities.

Digital divide

The emergence of a *digital divide*[3] began to be noticed in recent decades as a trend towards the increase of inequalities when going from a postindustrial society to the stage of technologically advanced societies. An initial approximation used the term *technological divide*, or the socioeconomic difference between those communities that had access to the Internet and those that did not; this was then extended to the concept of *digital divide*, where these inequalities were considered to also involve all new ICTs, such as the personal computer, the mobile phone, broadband, and other new devices.[4]

The repercussions of this issue in the international system were registered in the '80s, with the presentation of the International Telecommunication Union (ITU) report "The Missing Link" in 1985, where countries' unequal access to information technologies was described and the access gap to the (knowledge-intensive) emerging economic model, which was deepened with the inclusion of computers in developed countries, was made apparent. The ITU warned throughout the '90s that an informational and technological gap between countries was being promoted, which in turn was creating *information poverty* and, at the same time, a division between the *info-rich* and the *info-poor*. In this way, the connectivity asymmetries observed between different regions of the world, as well as the differences in access to new technologies that took place inside the same country, began to be tackled.

Currently, only 63% of the world population has access to the Internet and, at the same time, this connection to the virtual world is distributed in a very unequal way. While in North America and Europe, Internet penetration reaches 90% and 87% of the population respectively, in Africa it falls to 47% and in Asia it is 59%. While Europe and North America jointly have over 1 billion users, in Africa there are over half a billion, and these differences also appear between men and women, city or countryside, age groups, social status, and so on. (Internet World Stats, 2020).

---

[3] The origin of the term *digital divide* can be traced back to some internal US documents (1995 Department of Commerce Report titled "The Digital Divide, a Survey of the Have Nots in Rural and Urban America"), where the concept specifically refers to the recognition of a problem concerning the access to new technologies that occurred in the interior of the country and that affected specific collectives in specific places in the North American geography, which they called "internal digital divide."

[4] The OECD expanded the definition of digital divide as follows: "The gap between individuals, households, businesses and geographic areas at different socio-economic levels with regard both to their opportunities to access information and communication technologies (ICTs) and to their use of the Internet for a wide variety of activities" (OECD, 2001).

Cognitive divide

The evolution in the analysis of the *digital divide* has led to the emergence of new concepts that try to better explain the complex reality of the distribution of STI in the international system. In the scenario of *knowledge society*, the phenomenon of knowledge inequalities is called *cognitive divide*. These differences are notable on an international scale, where central countries are benefited from the production of scientific knowledge, whereas peripheral ones do not have access to primordial cognitive goods in important issues such as new medical or agricultural knowledge or educational material. The knowledge divide is evident between Northern and Southern hemisphere countries, but it also manifests itself internally in each society (UNESCO, 2005). Some experts suggest that the digital divide is changing from a gap in the access and connectivity to new technologies to a knowledge divide, which means the gap has gone beyond the access and having the resources to connect to new technologies, to the interpretation and comprehension of the information once connected.

For UNESCO (2005), the digital divide feeds the cognitive divide, which is much more concerning, as it accrues the effects of the different gaps observed in the main constituent areas of knowledge (access to information, education, scientific research, and cultural and linguistic diversity) and is the true challenge for the creation of knowledge societies.

Scientific divide

Lastly, the concept of *scientific divide* is linked with the increasing differences established between the different actors of the international system concerning R&D investments, the number of researchers, the production of scientific publications, patents, and the development of innovation processes (UNESCO (2005). The idea of a scientific divide does not only refer back to the existence of economic inequality but also to the divergences that affect the political conceptions of the social and economic functions of science. The scientific gap exists from the moment in which the governing powers do not consider science and technology as a primary economic and human investment.

The emergence of this divide has led UNESCO to wonder if the world is heading towards a dissociated society where knowledge is distributed unequally. "There is a real scientific divide setting the 'science-rich countries' apart from the others. Science is by nature universal, but scientific progress seems confined to one part of the planet only. A number of regions suffer from a considerable handicap, which is impeding the development of research" (UNESCO, 2005).

Impact of the divides

These knowledge divides, understood as new emerging realities within ISR, are having huge consequences in the international system. This impact is characterized by three features: (i) the generation of new knowledge centers and peripheries, (ii) the emergence of intrasocietal divisions, and (iii) the strengthening of old social divides.

New centers and peripheries

One of the direct consequences of the emergence of new knowledge divides is the creation of what can be understood as new *centers* and *peripheries* of scientific knowledge. The new realities of the emerging gaps increase the territorial differentiation between those regions that benefit from the production, intermediation, and application of scientific knowledge and those that are increasingly deprived of them.

Analysts such as Saskia Sassen (2007, 2017) and Daniel Innerarity (2011, 2013) underline the emergence of these new realities of centrality and marginality of scientific knowledge. Sassen considers that the technological revolution and the expansion of the global economy have contributed to generating a *new geography of centrality and marginality*, which links knowledge with particular *strategic places*, as STI ends up being distributed in centers and nodes that generate larger differences between countries and regions. The new geography of economic globalization simultaneously contains dispersion and centralization dynamics. Innerarity equally shows concern for the future role of knowledge and the geographical spaces it inhabits, understanding that information can be considered as universally accessible, while *knowledge* is linked to certain very specific territorial contexts: "While the industrial economy is delocalized and benefits from global reorganization, the creative economy tends to territorialize, to choose spaces conducive to networking and exchange." For this, Innerarity insists that it is necessary to ask for the real distribution of knowledge, the technological divide, the digital divide, and new regional peripheries. It is a new geography of centrality and marginality where the concepts of center and periphery must be rethought.

Intrasocietal differences

The second visible consequence of new knowledge divides emerging in the field of ISR is closely linked to the impact these divides have within Nation-States, as the scientific activities and knowledge are not only distributed unequally among nations but also within them, generating important intrasocietal divides. These divides are strongly increasing in all societies as a consequence of the high degree of concentration of research in specific places: in the United States, in 2016, more than half of its R&D expenditure was concentrated in five states, of which California represented more than one-fifth; Moscow represents 50% of Russian research papers; Tehran, Prague, Budapest, and Buenos Aires represent almost 40% of their national productions; and London, Beijing, Paris, and São Paulo are responsible for over 20% in their respective countries (NSF, 2016).

The divides, while also increasing between rich and poor countries, are growing within the rich countries' own societies, where between 25% and 30% of the population does not have access to computers or the Internet and, in places where physical access (digital divide) is widespread, there are problems of qualification and use (cognitive divide) (Van Dijk, 2020). Experts such as Castells, Sassen, and Knight understand that the great majority of the world's innovation centers are nodes (cities) which, through the connection with universities and certain public policies, are transformed into true poles of attraction for people and investment. Inevitably, the growth of those poles deepens the differences with the rest of the surrounding territories.

Reinforcing old social divides

Lastly, the distribution of scientific knowledge in the international system has also meant the rekindling of old social divides and their coexistence with new social, cognitive, and scientific divides. New inequalities in the access or use of new digital and knowledge technologies have started to join the traditional global inequalities (*social gaps*), linked with the level of economic income, access to healthcare and education, infant mortality, age, sex, or ethnicity. In this context, the old and new inequalities are overlapping, connecting, and feeding back into each other (Van Dijk, 2020). In short, the new gaps in scientific knowledge generate new inequalities, while, at the same time, reproduce and deepen old social differences, creating a vicious circle where the divides feed and feed back into each other.

6.3 Multilevel and Global Governance

The second emerging phenomenon in the current structure of ISR is related to the way in which international actors, through their interactions and interrelations, are shaping a new type of governance of STI, mainly characterized by being multilevel and global.

Currently, the processes of planning, organizing, and executing STI increasingly depend on the new phenomenon of governance. Now, international actors interact among themselves through links of negotiation, cooperation, agreement, information manipulation, and alliance creation to plan and execute actions linked to scientific knowledge. This introduces an important element of uncertainty in their interactions, as the relations between all these actors are no longer based on the hierarchical structures of a *government* (based on coercion, command, and control), but rather *on the emergence of a governance*, where the processes of relation and coordination between actors are more relevant.

In recent decades, these mechanisms have transformed as a consequence of the increase in actors taking part in these processes and of the acceleration and intensification of the interactions established between them. It is a change from a differentiated, hierarchical, and independent way of governing and managing scientific knowledge to a more numerous, complex, and interdependent governance carried out in multiple levels of action.

Multilevel governance

The first element to be considered from the emerging reality of scientific knowledge governance is linked to its multilevel character. Currently, the new scientific knowledge governance is manifested as a complex negotiation mechanism between multiple actors that are found in numerous territorial arenas or scenarios (on a local, national, regional, and international level). This has been part of an extensive process of institutional creation and decisional reassignment that has promoted some previously centralized functions of the State to the supranational level and some towards the local level (Piattoni, 2010, 2018).

It is a new type of government that involves public and private actors in the political process with a higher degree of cooperation. In its most general sense, multilevel governance evokes decision-making and vertically open and horizontally integrated public policy processes. Various actors, which develop links that are no longer hierarchical and coercive, but rather processes where horizontal interactions and interrelations becomes a key element, take part in those processes.

Generally speaking, the new multilevel governance of STI has four main traits:

– States no longer monopolize the creation of public policies or the aggregation of domestic interests; on the contrary, a large number of non-State actors have a special interest in the creation of scientific production policies.

- The decision-making process is shared by actors on different levels, owing to the strong interrelation between the agents that take part in different arenas and fields, who are now forced to cooperate in order to reach a consensus.
- Supranational institutions have more influence due to the increasingly important role these agents are playing on a global level.
- Subnational actors are not limited to their State or local field, but they operate in the State and global arenas creating transnational links (Hooghe and Marks, 2001, 2020).

Global governance

At the same time, the governance of STI is also showing signs of being more global. In this sense, it is not about considering the existence of a global government that regulates knowledge or the appearance of a new hierarchic top-down authority structure; on the contrary, this global governance must be considered as a collection of many levels of government related to the activities, regulations, and mechanisms, both formal and informal and both public and private, that exist nowadays (The Commission on Global Governance, 1995).

Global governance has resulted in new STI planning and organization mechanisms, where multiple actors work in an interconnected and interdependent fashion, creating new types of global management. Among these new mechanisms are (i) new IGOs and NGOs; (ii) international regulations, laws, and standards; (iii) international framework agreements and resolutions; (iv) international regimes; (v) *ad hoc* agreements and global conferences; and (vi) private and private–public hybrid governance systems (The Commission on Global Governance, 1995).

The two main elements that have stimulated and intensified global governance have been the increase in the number of actors and the growing number of interactions between them.

- *Plurality of actors*: A part of this phenomenon's complexity is due to the multiplicity of actors taking part in the creation, coordination, and execution of scientific knowledge. What is noteworthy about this new emerging reality is that all international actors are relevant and take part almost without establishing predetermined hierarchies. For this reason, there is a long list of actors (including Nation-States, sub-State entities, universities, IGOs, NGOs, transnational actor networks, epistemic communities, transnational companies, and think tanks, among many others) that play an active role in the new type of knowledge governance.

- *Multiple linkages*: This governance implies the interrelation of multiple social actors through links of cooperation, competency, agreement, and alliance making (Mariscal, 2006, 2011, 2017). The majority of actors take part in activities, initiatives, and dynamics that represent a certain degree of scientific knowledge governance through different types of association between scientists, governments, companies, and organizations that are being established to face science's global challenges. The previous, almost exclusive participation of State actors has given rise to various interactions and relationship dynamics that involve the participation of a large group of actors in both formal and informal initiatives. The agents that take part in these new dynamics generate new linkage structures, whose main objective is to carry out activities and cooperative agreements leaning towards problem resolution and common affairs.

As a consequence of the increase in the number of actors and interactions, the new governance of scientific knowledge must be understood as a new emerging phenomenon, mainly characterized by being *polycentric* (Todt, 2021; Innerarity, 2011), where the creation of policies is structured in complex networks through which different actors, relatively autonomous, but at the same time interdependent, interact.

Future challenges

The future governance of STI must face the great challenge of improving the levels of planning, organization, and communication between actors with the express objective of giving a more effective solution to the countless challenge it will face in the current international context. Governance will have to tackle the following issues:

- Reaffirming the need for scientific knowledge being understood and used as a tool for solving the *global challenges* humanity is currently facing, which transcend national borders and represent a significant threat to the international system.
- Considering the processes of scientific knowledge governance as the most adequate method for channeling, planning, organizing, and applying STI in the global context.
- Extending the information on the process of governance, as well as its main results, to the entire society, through effective mass media coverage of scientific research and its results, which will increase public awareness on the subject.

- Openly engaging in debate with regards to intellectual property and the international patent system, which have many issues and are hard to harmonize due to the multiplicity of interests at stake.
- Considering the open challenge posed to the international system by the increasingly unequal distribution of scientific knowledge and the increase of the social fractures that the knowledge society is creating.
- Democratizing the processes linked to STI by incorporating new actors, mechanisms, social sectors, and geographical spheres.
- Debating ethical and social solutions against the uncertainty generated by the consequences of new emerging technologies.

The global challenges of the international agenda, along with the problems of STI, foretell the existence of a structure and configuration of ISR that will be very hard to govern, which will require a process of large-scale international cooperation to tackle the nature and magnitude of the issues to be solved. In this international context, it is evident that no international actor on its own will be able to offer definitive solutions, which is the largest challenge for the multilevel and global governance of STI.

6.4 Commercialization of Science, Technology, and Innovation

Another of the realities emerging from the analysis of ISR is linked with the increasing commercialization and marketing of scientific knowledge in the international system. The study of the context, the actors, the relations, and the main processes of scientific knowledge (production, intermediation, distribution, application, and governance) has allowed us to discover a great commercial interest in STI. This increase is explained by the economic reevaluation of scientific knowledge and by the intensification of the competition between the actors of the international system, which has caused an unprecedented expansion of the market for those goods, services, and processes derived from STI. In the framework of a *knowledge economy*, it is not strange that many actors are prioritizing the economic value of scientific knowledge over its other uses, which has a great impact on the international system as a whole and, at the same time, a strong social impact.

Commercial value of science, technology, and innovation

Driving the commercialization of scientific knowledge is the intensification of international competition in business and industry, which is now spilling over into the pursuit of innovation. It is no longer enough to create new knowledge, now that knowledge must generate technological developments and result in innovations. Scientific knowledge by itself is no longer enough, and scientists need to not only disseminate knowledge in the field of their own peer community but also transfer it to other agents that can give it another use. Considering that the current capitalist economic system is characterized by the superiority of innovation over scientific research and technological developments, what is required is to transfer knowledge to companies, as they are the ones who generate innovations and, in turn, economic development and social well-being.

Following Javier Echeverría's (2008) analysis, the process of knowledge transfer can be understood as a *value chain* with many links. Scientists are the first link, as they provide basic knowledge validated by scientific communities. Afterward, this same knowledge must be transferred to the other links in the chain, where it is processed, manipulated, and transformed to obtain innovations from it. In this new phase, other people, less focused on theory and academic prestige and more focused on practice and the generation of economic benefits, transform knowledge into valuable innovations that are later protected through patents, use licenses, and juridical contracts.

A key moment takes place between these stages when the traditional development of science becomes a marketable innovation. Through this process, what is questioned is the value of classical science, that is, knowledge understood as a means unto itself.[5] In fact, it is not denied that it is an asset, but it is not the main one, which means, in short, the subordination of science to economic, political, and military objectives as one of the essential characteristics of contemporary scientific activity (Echeverría, 2014).

Economic models of knowledge production

Currently, the majority of methods of scientific knowledge production privilege its economic and commercial uses over any others. The majority of models understand (in an almost linear way) that knowledge production methods and all accompanying processes must be directed to the promotion of STI as a motor of economic development and as a first step towards the eventual progress and general societal well-being. Currently, there is a wide

---

[5] Gibbons et al. differentiate between "knowledge-based industries" and "knowledge industries." The former understands and improves the functioning of the production process and care about the product and its development process; the latter only trade in knowledge (Gibbons et al., 1994; Nowotny, Scott, and Gibbons, 2003, 2006).

consensus in the sense that a country's wealth and development possibilities are linked to the consolidation of STI sectors, to the critical mass of scientists and professionals linked to the productive sector, to the generation of research and innovations, and to the push for new emerging technologies in the most dynamic economic sectors. In short, underneath this perspective is the idea of producing STI directed at economic and commercial usage as a cornerstone for economic and social development, and the majority of the currently predominant scientific knowledge production models (*Innovation systems, Mode 2, Mode 3, Triple Helix, Quadruple and Quintuple Helix*, etc.) are based on this premise.

Decline of societal relevance

Some experts question the widespread idea of the production of scientific knowledge that ends up generating uses of STI that are tightly linked with the demands of the market. In essence, they argue that, if the forces controlling knowledge creation are the market and commercial interests (adopting the shapes of science commercialization or academic capitalism), we should have low expectations for that production and its applications and uses for a general and sustainable social development (Bourdieu, 2004; Núñez Jover, 2006, 2020; de Sousa Santos 2005, 2015, 2018; de Sousa Santos and Meneses, 2019).

The main argument is that the articulation between university, research, and economic sectors does not mean directly following the needs of the large majority. As per Pierre Bourdieu (2004), the consequences of this commercialization of STI begin to show serious consequences: "The pressures on the economy are increasingly daunting, especially in those fields where the results of research are highly profitable, such as medicine, biotechnology and, more generally, genetics, not to mention military research […] This is how both researchers and research groups fall under the control of great industrial firms dedicated to maintaining the monopoly of highly-profitable products through patents."

The increase in the commercialization of STI puts into question the application, uses, and utility of knowledge within ISR, and opens a serious debate on the impact and societal relevance of the scientific knowledge produced. The main question is whether the consequences of the production, intermediation, distribution, application, and governance of scientific knowledge are considered positive or negative according to society as a whole, insofar as they allow this new scientific and technological knowledge to be incorporated and used by a large majority of citizens.

6.5 Gender Divide

Traditionally, the role of women in the field of STI has been always subordinate to that of men, and, despite having made important contributions to science throughout history, in the majority of cases, their participation has been forgotten or, on multiple occasions, even hidden. Only in scant occasions has a woman's contribution been remembered and recognized, such as with Marie Curie or Rosalind Franklin. This anomaly is mainly based on the prominence of a patriarchal system throughout the international concert, by which women have had limited access to higher education and the most relevant job positions when they were not outright denied their place. This discrimination has had marked repercussions in the field of STI. This paradigm has started to very slowly shift since the beginning of the 20th century, gaining speed in the second half of the century as a result of significant demographic and cultural transformations concerning the role that women occupy (or should occupy) in society. These changes, although limited, are having important social, political, economic, and cultural consequences for the entire society in the 21st century.

Nowadays, many experts wonder about the scientific, economic, and social impact that removing all social and cultural barriers that limit woman's active participation in science and incorporating them on equal footing with men would have. The answers are always positive: not only experts but also many international organizations and governments are recognizing that gender equality in the field of STI would open a new horizon in the field of scientific research, allow us to advance in new scientific solutions, and end in greater economic and social development.

Statistical proof shows that, although it is possible to identify a greater importance of women in the field of STI, this growth is not sufficient to reach true gender equality, as the majority of barriers limiting the full participation of women in science have not been removed.

Women in science, technology, and innovation

The field of higher education has been the domain where women have made the most progress in recent years, even becoming the majority in many countries and areas of knowledge. Generally speaking, women represent 53% of total graduate and postgraduate students on a global level (UNESCO, 2015). Currently, women are the majority of graduates in four fields of higher education: humanities and arts; education; social sciences, business, and law; and health and welfare; on the contrary, in some technical areas such as natural sciences, engineering, and ICTs, women are still a minority. In the OECD, only 30% of graduates in these areas are women (OECD, 2017).

The most dramatic change seems to be in the move towards doctoral studies and, fundamentally, the jump to research careers and other positions linked to science where women are markedly underrepresented. Sophia Huyer calls this situation a "leaky pipeline," a phenomenon by which women reduce their participation in STI fields as they climb up the professional ladder: from the 53% of women in undergraduate and postgraduate programs, only 43% continue their doctorate studies, and they represent 28% of all researchers (UNESCO, 2015). This reality seems to indicate that the majority presence of women in higher education does not translate into similar representation in the field of scientific research.

On an international level, women represent 25% of all researchers, although there are very notable variations depending on the country and the region. Regionally speaking, Central Asia (47%), Latin America and the Caribbean (45%), and the Arabic countries (40%) represent the best indicators, while Asia shows the lowest global average for a continent (20%). It is interesting to point out that many developed countries have a low proportion of female researchers, such as France, Germany, or the Netherlands (only 25%). The percentage is even lower in developed Asian countries such as South Korea (18%) and Japan (15%) (UNESCO, 2017).

Similar results can be found in other scientific areas where women are underrepresented: only 22% of scientific projects are carried out by women; this number is even lower for subgroups of writers, such as those engaged in peer review or publishing activity, or those who are full-time researchers; the percentage of patents presented by female inventors varies from 4% in Austria to 15% in Portugal; in public and private universities the proportion of female rectors was 20% on average in the European Union in 2014; lastly, it is noted that in scientific and academic work, women often earn considerably less than men (OECD, 2017).

Ambivalent trends

The statistical analysis of women's participation in STI seems to confirm at least two clear trends within the field of ISR: (i) the evolution and growth of women in the field of higher education, in new disciplinary fields, and also in authority positions in scientific spheres is evident; however, (ii) this growth is not significant enough to break with male historical prominence, which causes the persistence of the gender divide in the field of ISR.

The explanations behind this persistent gender divide are multiple and include economic, social, and cultural variables:

– On the one hand, it is understood that highly qualified working women are exposed to invisible barriers that block them from reaching the highest

hierarchical levels, independently of their achievements and merits (*glass ceiling*).
– Women also face limitations that impose on their tasks such as taking care of the family, to which they have been traditionally relegated and which prevents them from achieving complete professional development (*sticky floor*).
– Finally, the multiple cultural limitations, linked to patriarchal and paternalist systems that reproduce women's subordinate roles, to which women are subjected throughout the international system must also be considered.

The permanence of all these barriers has driven voices claiming the need for "fixing the system" (UNESCO, 2015) as the only possible solution to achieve a real gender equality through policies and strategic actions that include all actors of the international system.

Public policies

Although in recent decades, many public policies have been designed and applied throughout the international system to stimulate the participation of women in science and reduce the gender divide, none of these programs and public and private strategies have been enough to remove the tangible and intangible barriers that limit women's full participation in the field of STI.

The lack of results or the direct failure of many of these policies has pushed many actors, such as the European Union, to revise and fix their own strategies. Beyond the measures introduced in the Horizon 2020 Research and Innovation Framework Program, the European Union has decided to double its commitment to women through the ERA Roadmap 2015–20, which establishes "Gender Equality and Mainstreaming in Research" among its priorities. At the same time, the European Parliament adopted, on September 9, 2015, a resolution on women's career paths in the scientific and academic fields, offering incentives for research centers and universities to design and apply gender equality plans, to incorporate the dimension of gender in their national research programs, to eliminate legal and other barriers for the hiring, retention, and career development of female researchers, and to implement global structural change strategies to fix existing gender inequalities between the institutions' professors and authority figures and in research programs.

None of the previously implemented public policies linked with the promotion of women have had an astounding success; however, without these efforts, it would seem very unlikely that the system would end up intrinsically stimulating gender equality.

6.6 Geopolitical Changes

Some of the most relevant phenomena for international relations are the geopolitical and geoeconomic changes that can take place in the world order. For this reason, one of the most significant emerging realities that can be observed after analyzing ISR is the displacement of the focal points of STI from Western countries towards Asia. It is not a sudden change, nor is it an absolute loss of power for traditional Western actors, but rather a slow but persistent displacement towards Asian countries, which are having an increasingly decisive weight in the current global structure of ISR.

Historically, science has been understood as a modern enterprise, where the United States and some nations from Western Europe have maintained a prominent position as a consequence of the strong investments that have been made in their educational and scientific and technological systems. However, in recent years, many other States have started to recognize the importance and the benefits (both economic and social) of investing in science and have developed ambitious R&D investment plans seeking to develop new research infrastructures to promote internal talent and incentivize the attraction of foreign talent. All emerging countries, starting with the so-called BRICS, as well as countries with low economic development, have begun to seriously invest in STI. These new plans by State actors who have not traditionally taken part in ISR have started to cause significant changes in the international system and to show a very significant trend of geopolitical displacement in the production of scientific knowledge.

Regional changes

The statistical analysis carried out in the field of ISR has allowed us to identify multiple considerable regional changes linked to STI that put Asia in the spotlight. Among the most significant are as follows: (i) East Asia[6] has become the region with the largest R&D investment in the world (44% in 2020); (ii) Asia has consolidated its leading position in the number of researchers, growing from 35% (2002) to 43% (2013) of the world total (China, 19%); (iii) Asia has become the region with the largest quantity of scientific publications (39.5%), surpassing Europe for the first time in history in 2014; and, lastly, (iv) Asia is the region with the largest number of patents in the world (47.5% in 2016).

---

[6] Formed by the following countries: China, Taiwan, Japan, South Korea, Singapore, Malaysia, Thailand, Indonesia, Philippines, Vietnam, India, Pakistan, Nepal, and Sri Lanka.

The majority of Asian countries have shown unparalleled growth and expansion in all indicators linked to STI: R&D investment, number of researchers, publications, and patents. With China at the lead and India and South Korea following close by, Asia has grown at a greatly superior rhythm than that shown by traditional STI actors such as the United States and the European Union. The majority of Asian countries have increased their yearly R&D by rates between 7% and 10%, while the rest of the actors in the international system did grow by 2% to 4% per year. Scientific publications grew particularly in China, Singapore, and South Korea, while the USPTO patents also rapidly increased owing to Taiwan, South Korea, and China.

All this data allows us to firmly verify the rise of Asia, as the most important region in the field of ISR at the beginning of the 21st century.

China's rise

Within the generalized and extraordinary development carried out in STI in Asia, China is particularly noteworthy. Although all the regions have grown, and it is possible to find other interesting examples such as South Korea, Singapore, or Taiwan, without a doubt, the paradigmatic case is China. In the last two decades, China has made a strong and decided commitment in the field of STI, which has resulted in a sustained development that has turned it into one of the most important actors in 21st-century ISR.

The evolution of its STI system has been incredible in the last few years, and it is enough to observe just a few indicators to verify its rise: (i) the average growth in R&D was 20.5% per year between 2000 and 2010, and 14% between 2010 and 2015; (ii) it is the country with the largest number of scientific publications in the world (18.6%), overtaking, for the first time in history, the United States in 2017 (17.8%); (iii) the percentage of its GDP invested in R&D reached 2.09% in 2014, which allowed it to pass the European Union's average expenditure; (iv) it has become the nation with the largest quantity of researchers in absolute numbers with 1,318,000 in 2013 (approximately 19% of the world total), first surpassing the United States in 2011, and very close to reaching the European Union; and, finally, (v) in 2019, it was the first year since the PCT system began operating in 1978 that applicants residing in China filed the most applications (58,990 PCT applications), moving down the United States to the second place (57,840 PCT applications filed).

The STI development of China in the last two decades is an amazing fast-growing phenomenon. In 2000, China accounted for nearly 5% of global R&D, joining the United States, Japan, South Korea, and the countries of Western Europe as the largest funders of R&D. In 2009, China surpassed Japan to become the second largest funder of R&D. From 2000 to 2018, while

China's share of global R&D rose from 4.9% to 26.3%, the US share fell from 39.8% to 27.6% and Japan's share fell from 14.6% to 8.1% (CRS, 2020).

In the race for emerging technologies, China also shows an extraordinary performance comparable only to the United States or the European Union. For a few years, China has been strongly investing not only in *artificial intelligence*, considering it the key element for the future world economy but also in areas such as *biotechnology*, *robotics*, and *precision medicine* (WEF, 2018). These investments have allowed it to grow in 2018 with record-breaking numbers in multiple areas: the high-tech sector increased by 12.5%, robot production expanded by 35.4%, integrated circuit (high-tech chip) production increased by 17.2%; and electric vehicle production increased by 56.7% (Castro, 2018).

Geopolitical and geoeconomic shift

The main trend emerging from this statistical analysis is not only an extraordinary growth in Asia (especially Southeast Asia and China) but also the realization that this development happened at the expense of the United States and Europe, which allows us to elucidate that what ISR is currently experiencing is a geopolitical and geoeconomic shift from the West (United States and European Union) to East Asia (led by China) in the field of STI.

This geopolitical and geoeconomic shift is evident when each of the most relevant dimensions linked to STI is analyzed:

- East Asia shows high and constant growth in its R&D investments, which has allowed it to rise from 25% of the global total (2000) to 44% (2018), and which took place at the expense of North America's decline, which decreased from 40% (2000) to 25% (2018), and Europe's, whose global presence fell from 27% to 20% in the same period.
- China represents a third of the world's growth in R&D investments between 2000 and 2018, approximately as much as the United States and the European Union combined.
- Asia's share of the world's researchers has grown from 35% (2002) to 43% (2013), while, in the same period, the United States decreased from 23% to 16% and Europe from 33% to 30%.
- Lastly, patent registrations in Asia (mainly China, Japan, and South Korea) grew from 18% (2002) to 47.4% of the world total (2016). In this same period, both Europe and the United States decreased by 14%.

Despite average annual growth in R&D spending of 4.3% in the United States and 5.1% in the European Union between 2000 and 2017, global R&D shares declined for the United States (37% to 25%) and for the European

Union (25% to 20%). At the same time, the economies of East-Southeast and South Asia—including China, Japan, Malaysia, Singapore, South Korea, Taiwan, and India—increased their combined global share from 25% to 42%, so this region now exceeds the respective R&D shares of the United States and European Union and leads in global R&D expenditures (National Science Board and National Science Foundation, 2020).

Despite the United States' and European Union's great efforts to prioritize investment in STI, and despite them maintaining a privileged position in ISR, the data analyzed here shows three substantial geopolitical changes:

(i) East Asia's superlative growth led by China.
(ii) A relative decrease in Western countries, mainly the United States and the European Union.
(iii) A slow but constant geopolitical shift in STI from the West to East.

These changes mean the creation of a new scenario in ISR that will have a strong impact on the way international actors polarize and balance the international global system.

6.7 Science and Emerging Technologies

An incredibly important element in the analysis of ISR is the emergence of new areas in science and technology. These areas, which have appeared as a result of the evolution and development of scientific knowledge, are having more and more relevance in the scientific and technological fields and, especially, a considerable impact on the application of the new knowledge in the economic, social, military, and political domains.

Science has advanced considerably throughout history, creating new scientific knowledge on top of previous knowledge. This idea, attributed to Newton, considers science as the human enterprise that allows the discovery and knowledge of new facts, realities, and circumstances in a systematic and reliable way. It is a process that offers the possibility of expanding our cognitive horizons and the frontier of human knowledge. Throughout the last decades, multiple factors have allowed science to rapidly advance through new scientific and technological discoveries and inventions that open the gate not only to new knowledge but also to innovative applications. As has happened in other moments of history, but now with a much greater intensity, this new context also means tackling uncertain future scenery, due to the unforeseeable consequences and social, political, and cultural impacts these new discoveries may have.

The acceleration of scientific and technological advances in recent years are the product of several factors that have jointly encouraged it: more interest in the creation of scientific knowledge, more actors involved, more STI plans and investments, more researchers and highly skilled workers, new and varied knowledge production methods, more intense cooperation, more applications, and more demands from multiple sectors, among others.

New scientific and technological areas

New scientific and technological areas include a wide variety of sectors and fields such as *educational technology, information technology, nanotechnology, biotechnology, cognitive science, psycho-technology, robotics, and artificial intelligence*. Schwab (2018) points out the emergence of new areas of science application such as *artificial intelligence*, the *Internet of Things* (IoT), *autonomous vehicles, 3D printing*, or *precision medicine*, that promise to expand people's and humanity's possibilities in extraordinary ways. The key of these new scientific and technical fields is their capacity of becoming disruptive and innovating factors that can radically change an industry, or even create new ones.[7]

Among the new areas and fields of knowledge that have strongly developed in the last few years and that have a strong disruptive potential, we find the following:

- *Robotics*: It is the design, construction, operation, manufacture, and application of robots that combine a great diversity of disciplines such as mechanics, electronics, information technology, artificial intelligence, control engineering, and physics. In recent years, robotics has exhibited great development and promises a significant reduction in cost, an increase in productivity and in the quality of products, services, and processes, as well as applications in numerous fields.
- *Biotechnology*: It is based on a mix of engineering, physics, chemistry, medicine, veterinary medicine, and agricultural sciences. It has great promise in the fields of agriculture, industry, medicine, and health, in the search for new energy sources and the creation of new materials.
- The *Bionic* industry: It links the science of living organisms with modern architecture, design, engineering, and modern technology, opens unexpected possibilities for living beings to improve in many ways thanks to the help of mechanical tools.
- *Nanotechnology*: It is essentially the precise manipulation of atoms and molecules for the creation of product on a microscale and combines fields

---

[7] The concept of *disruptive innovation* was introduced by Clayton Christensen in 1997 in his book "The Innovator's Dilemma," representing how a product or service, which originally appears as something residual or as a simple application without many followers or users, can quickly become in the leading product or service in the market.

such as surface science, organic chemistry, molecular biology, and physics, among others. Consequently, nanotechnology could be used to create new materials and devices with a wide range of applications, such as in medicine, electronics, biomaterials, and energy production.
- *Artificial intelligence*: It is understood as machines' and robots' capacity for learning and solving problems by themselves. Although the origin of this field of computer science is traced back to 1956, when the term was coined by John McCarthy, nowadays its horizons have been expanded as new ICTs allow computerized systems to process large amounts of data and increase the speed, size, and variety of information that can be gathered. Currently, artificial intelligence can carry out tasks such as recognizing data patterns in a faster and more effective way than humans can and making more effective decisions through algorithms.

Following Schwab's (2018) observations, the emergence of these new scientific and technological areas opens a great range of possibilities for its application in different fields of human life:

(i) The *expansion of digital technology* with the development of new information technologies, higher storage capacities, and the advance of the IoT.
(ii) *Physical-world reforms* through the rise of artificial intelligence, robotics, the usage of drones, the discovery of new materials, the development of autonomous vehicles, or 3D printing.
(iii) *Changes in human nature* through biotechnology, synthetic biology, or neurotechnology.
(iv) *Environmental integration* with new sources of energy, forms of storing and transmitting it, and the development of space technology.

Future implications

The emergence of new scientific and technological means the possibility of discovering and producing new knowledge with applications and usages in a wide variety of areas which, at the same time, will inevitably open an unknown domain with regards to the social, political, cultural, and economic consequences that these new technologies will generate. According to Schwab (2018), "these new technologies have the potential to change the course of history and affect every aspect of our lives." In the same line of thought, Joe Kaeser (2018) points out that scientific and technological changes will transform all human activities: the ways in which we use natural resources, communicate and interact, learn, work, govern ourselves and do business, among many other changes.

Although the implications of these advances are still largely unknown, at the beginning of the 21st century we can already identify some of the main consequences and challenges these new areas of science and technology are generating in different human areas:

- *Scientific*: The new areas in science and technology are creating a true revolution in the scientific field through great changes in the way science is done. Even if, in Thomas Kuhn's (1962) terms, *anomalies, paradigm,* and *model* changes and, ultimately, *scientific revolutions* are a substantial part of science's progress, nowadays we seem to be facing a scene of radical change in scientific research, a result of the emergence of new scientific and technological fields. Among the most visible modifications that can already be observed are the facts that (i) traditional disciplines are readapting their theoretical, methodological, and cognitive frameworks to the new empirical realities; (ii) new disciplines and subdisciplines that address these new scientific areas are appearing; (iii) the number of approaches involving more than one disciplines is increasing, which is stimulating new *inter-* and *transdisciplinary* fields through the increase of cooperation between disciplines; (iv) new techniques, methods, and methodological frameworks are used as technology open up new opportunities; (v) new actors are involved in the creation of scientific knowledge, increasing democratic participation but, at the same time, increasing its complexity.
- *Economic*: Scientific production, previously a characteristic of laboratories and research areas, is now closer to productive sectors due to the economic relevance of these new scientific and technological areas that are creating new business fields and larger productivity margins. The possibilities offered by the virtual world, robotics, or artificial intelligence are still unquantifiable for the economic system, but what is clear is that they will have a great impact on economic processes, which foretells significant changes in the methods and factors of production, consumers' preferences, the qualifications required by the workforce the way of promoting and selling products and services, and the places where the main markets are concentrated, among others. In the prelude of great changes in the economic system, the main international actors (not only companies) are making great investment efforts in the emerging sciences and technologies with the objective of applying their results to the 21st century's new digital economy.
- *Social*: The creation of new scientific and technological knowledge is permeating virtually all sectors of society, creating a true revolution on the quantity, speed, and relevance of knowledge that can be handled by society. The development of new scientific and technical areas opens a great variety of applications for emerging technology in the social sphere,

which entertains the idea of being able to face humanity's great challenges but, at the same time, generates concerns as to the social distribution of these advances, considering the precedents of the previous industrial revolutions and the current unequal distribution of STI in the current international system. Additionally, changes in society will have great consequences on people's individuality, which opens a very wide spectrum for psychological responses towards such a variety of external changes.

- *Geopolitical*: The increase in the quantity and intensity of international actors (both State and non-State) interested in developing new scientific and technological areas as a tool for economic growth and social development and as a strategic factor of empowerment is already having a great impact in the geopolitical and geoeconomic distribution in the international system through the stimulus of new polarization and the appearance of new actors, which foreshadows a strong international dispute in the coming decades.
- *Military:* The new scientific and technological sectors are having a strong impact on military strategies and development, as more and more international conflicts are solved through the intensive use of digital media and in the *virtual world*, rather than the traditional military superiority, which is a true revolution in the field of international security and defense. In the same way that the intensive application of technology in weaponry (drones and autonomous vehicles) and soldiers (specialized uniforms) are changing the traditional way of fighting, *cyberspace* has, at the same time, become a new field of military operations, where technology is the main weapon and hackers are the best soldiers.
- *Ethical:* Advances in the field of science and technology have also led to the emergence of new ethical challenges that leave traditional authorities (whether State, scientific, religious, community, or civic) to develop new responses to guide the course of change and its ethical and social consequences. Many experts point out their concern with regards to what they believe to be an excessive and uncontrolled development of emerging technologies (N. Bostrom and J. Hughes)[8] and even some companies in the technological area, such as Google, show interest in better knowing the moral consequences of their work, creating special units to analyze, for example, the ethical challenges of artificial intelligence.[9] As Emilio Lamo

---

[8] Nick Bostrom and James Hughes founded the think tank "Institute for Ethics and Emerging Technologies" in 2004. Bostrom has also founded the Oxford Martin Program on the Impact of Future Technology, in the Oxford University in 2011.

[9] *DeepMind Ethics & Society* was created by Google in 2017 as a research unit on the ethical challenges faced by emerging technologies.

de Espinosa (1994) points out, "Scientific knowledge puts in our hands possibilities to do things for which we completely lack ethical guidelines."

6.8 Virtual World

One of the most notable and important emergences that stems from the analysis of IST of the beginning of the 21st century is the appearance of a new space of human relation and activity, called the *virtual world*. This new *cyberspace* is a simulated reality that is found inside computers and digital networks all around the world, which has become a new and very relevant terrain and environment for human interactions.

The emerging phenomenon of the virtual world appeared as a consequence of the development of innovating technologies in the field of information technologies, which have led to the improvement and the expansion of Internet and digital devices. This has created a virtual world, parallel to the physical one, which is having a substantial impact on practically every human and social activity.

Development and cooperation

The evolution of the virtual world in the 21st century has been a true revolution not only in the way humanity interacts but also in the way that it carries out its activities. Cyberspace is radically changing how we interact and socialize (social media), work (remote working), learn (rise of the MOOCs), organize our financial life (online banking), keep informed (digital media), buy (online shopping), and entertain ourselves (online videogames). New areas and human activities that were traditionally carried out in a specific physical place are increasingly moved to the virtual world; for example, medicine, religious worship, or contemplating art.

The development of the virtual world has allowed a faster and better connection between geographically distant people and has vigorously boosted cooperation and collaboration mechanisms all throughout the planet. Areas such as finance, business, and education are making use of the virtual world to operate more efficiently and productively, thanks to the advantages offered by immediate connection and communication as well as long-distance cooperation.

The virtual world shortens distances (ubiquity) and time (instantaneity) and opens a new, largely unexplored channel for the development of new ideas, expressions, links, and human activities. In recent years, the virtual world has grown and expanded exponentially owing to the continuous improvement and expansion of connectivity, electronic devices, and the development of technologies such as artificial intelligence or the IoT, which are increasing and improving the interconnection between the Internet and digital devices.

Virtual conflicts

With the rise of networks and the Internet, the virtual world has become a new space of human interaction where a large part of the economic, political, social, and cultural life has shifted and, consequently, also a large part of global bids and struggles. In parallel to social activities going virtual, the menaces to the network system have increased. It is not strange that, recently, cyberspace has been ravaged by the so-called *cybercrime*, which has made *cybersecurity* a core topic in our current international agenda.[10]

The key for this change is the appearance of new ways of aggression that are not only physical but also carried out in the virtual world. For international relations and geopolitics, this is a radical change, as it is now possible to fight an opponent by attacking the computer systems tasked with controlling the majority of their infrastructure and services through the Internet and networks. This means that international conflict has, to a large extent, shifted to cyberspace and the virtual world as a new operation field; it transforms information and communication technologies into new weapons and turns communication experts (hackers) into the new soldiers of the 21st century.

The consequences of cyberattacks are potentially devastating for any country, considering that, thanks to advances in science and technology, a large part of the security and defense, and the majority of their public services, are currently dependent on their computer structure. An accurate attack on a country's computer structure would completely paralyze its information, communication, and logistical systems; its defensive infrastructure; and its terrestrial and aerial transport operations, among other services. The potential consequences of a cyberattack against national objectives have led to the exponential growth, both in number and intensity, of these types of phenomena in recent years.[11]

---

[10] According to estimations by McAfee and CSIS (2018), the annual cost for the world economy of cybercrime is over 600 billion dollars (or 0.8% of the world GDP), and the trend is growing.

[11] In 2003, Taiwan received a cyberattack (attributed to China) that shut down much of its basic infrastructure. In 2007, Estonia blamed Russia for many attacks that altered the normality of media, banks, and government agencies. In the latter part of that same year, Iran also registered an attempted assault on their nuclear program through a program that infiltrated their computer networks (Stunex), which the Iranian regime has blamed on the United States and Israel. Lastly, one of the most renowned attacks has been the leak of e-mails of the United States Democratic Party in the 2016 elections. Investigations carried out by the United States government and intelligence

A context where connectivity is extended and expanded on a global scale, most of a country's essential infrastructure and services are computerized, and cybercrimes are easily committed, results in the certain possibility of being able to use new ICTs to inflict devastating damage on an enemy country. The idea of *cybernetic war* appears in this context, defined as "the actions of a Nation-State to penetrate into the computers or networks of another state with the goal of causing damages or distress" (Bankinter, 2016).

Cyberspace governance

Although the benefits of the virtual world are multiple and astonishing, the magnitude of the danger that the new *cyber conflicts* represent forces us to think of a cooperative governance between the actors of the international system that coordinates and regulates a large part of its uses and applications. Currently, the lack of legal frameworks for virtual space is allowing many States and non-State groups easy access to the destructive usage that virtual weapons can have. It is what Andrew Krepinevich (2011) calls "democratization of destruction." The tremendous damage that can be caused by cyber warfare, coupled with the ease with which it can be carried out, has in recent years fueled *cyberterrorism, cyberespionage*, and the *cybersecurity* concerns of many countries and companies. It is not strange to observe the budding race that governments and the private sector have started to recruit young programmers and obtain tools that would allow them to defend themselves from potential attacks. States have taken good account of the potentially devastating damage that can be caused by a cyberattack and have begun to plan methods of defense.[12]

Whatever the case, it now seems relevant to start thinking about the construction of a cyberspace order regulated by international actors that would avoid potentially devastating consequences for the international system. The fields of regulation, coordination, and cooperation in the virtual world are multiple and should consider topics such as ensuring the flow of information, the fight against cybercrime and cyberwar, the protection of privacy, the expansion and extension of connectivity, the pacific usage of networks, the development of digital economy, the boost to online education and learning, and the diffusion of culture.[13]

---

services blame the Russian government of using hackers to hamper Hillary Clinton in the race for the White House.

[12] Following estimates by Gartner, Inc., the cybersecurity business will be worth 150 billion euros worldwide by 2021.

[13] On March 1, 2017, the Chinese government published the document "International Strategy of Cooperation on Cyberspace," edited by the Ministry of Foreign Affairs and

All these changes caused by the emergence of the virtual world do not mean that military conflicts as we know them will disappear (at least in the short term); rather, it seems to suggest that we are in the middle of a transition to what Schwab (2016) calls "hybrid conflicts," where the physical and the virtual world coexist and constantly interact. In the new international context of hybrid conflicts, STI will play a leading role as it will be a strategic element in the development and application of technological advances for the armies and weaponry that will fight in the physical world, as well as being the main input for *virtual fighters*.

---

the Cyberspace Administration of China. This document is the first official proposal by the Chinese government on this topic.

Chapter 7
ROLES OF SCIENTIFIC KNOWLEDGE

The systemic analysis of the ISR (Part 2: Analytical Framework) has revealed that scientific knowledge has a privileged place in the 21st century's international system. Without a doubt, STI has been an important factor throughout history, but in the current world order, scientific knowledge has acquired an important role that transforms it into a key resource. For this same reason, the goal of this chapter is to offer a detailed and in-depth explanation of the roles and functions that knowledge has in the field of ISR and the influence that it has on the entire international system.

As a result of the observations throughout the investigation, it is possible to consider at least five main roles that scientific knowledge has in the current context of ISR:

– as a *resource for economic gain*,
– as an *instrument of power*,
– as a *mechanism of social innovation*,
– as a *democratizing element*, and, lastly,
– as a *military-strategic factor*.

7.1 Resource for Economic Gain

Nowadays, scientific knowledge is becoming an essential resource for the *generation of economic wealth* in the current context of the capitalist economic system. Although scientific knowledge has been used in this way in the economic field at least since the Industrial Revolution, what is novel now is its absolutely central and essential role in this new economic stage as a resource to obtain economic benefits.

In recent decades, scientific knowledge has made headway as a primary resource in the capitalist economic system. Since the end of the '60s and the beginning of the '70s, a new scenery of an international, postindustrial economy started to emerge, in which scientific knowledge became one of the most relevant resources for the generation of economic gain (Drucker, 1969;

Bell, 1974). Since the '80s, knowledge started to be considered as a key factor for growth, innovation, and the competitiveness of companies, which allowed for a fundamental change in the world economy, going from *comparative advantages* (David Ricardo, 1817) to the new importance of the *competitive advantages* (Porter, 1998, 2011). This evolution in the role of knowledge started to consolidate in the '90s when transnational companies started to focus their strategies on the more efficient creation, application, and expansion of *knowledge*.

What is new with regards to other historical moments is the relevance that STI currently has for the global economy, where the intellectual capital, producer of knowledge, is the most used factor of production for the innovation of economies. We are in an age where knowledge has become a key and distinctive factor that allows transforming inputs in goods and services with greater added value (García Guadilla, 2005, 2010). "The application of knowledge, which manifests as entrepreneurship and innovation, research and development, and product and software design, is the main source of growth for the world economy" (World Bank Institute, 2010).

The context of *knowledge economy* means the evolution from an economy based mainly on fabrication and industry to one based on scientific knowledge. Generally speaking, it is based on the use of ideas more than physical skills, or in the application of technology rather than on the transformation of raw materials or the exploitation of the workforce. Natural resources and their abundance (or scarcity) have lost a large part of their capacity to explain productivity and growth inequalities between countries and the economy's center of gravity has shifted towards the production of goods and services with information content, to technological innovation, and to the accelerated creation and productive use of new knowledge.

In the current international context, scientific knowledge acquires a special and strategic relevance, precisely because it has reached a significant economic usefulness, as the capitalist economic system's most dynamic sectors function thanks to the intensive use and innovative application of knowledge as the main source for the generation of economic wealth. On this foundation, the application and helpfulness of scientific knowledge is measured in economic terms, and it depends on commercialization and market logic (supply and demand). Any development from scientific knowledge, whether goods, services, or processes, that has commercial success will be seen as useful in terms of economic benefit and wealth generation. In the 21st century, scientific knowledge used in the production and management processes is definitely installed as a determining factor for the expansive dynamic and the competitive capacity of international actors, which turns it into a key piece in a country's or a company's economic-strategic actions as the main resource to sustain the generation of economic benefits and to boost growth and economic development.

7.2 Instrument of Power

A second and important role of scientific knowledge is the one *linking it with power* and with the idea that scientific knowledge gives power to those who sustain it and control it. The capacity to obtain and use power has been one of the main functions of any actor in the world order and, as such, a main element of study in international studies. Throughout history, there have been many factors that have allowed power to accrue as long as it was held: in some cases, it was cultural or religious justifications; in others, the threat or use of physical force through military apparatuses; in others, it was economic control.[1] What is new in the current context of ISR is the fact that STI has become one of the most relevant instruments of power in the international system.

Historically, knowledge has been associated with power in one way or another. With the advent of modernity and with the rise of scientific knowledge, the control of this resource of power became important for all the actors in the international system who wanted any prominence in the world order. From Ancient Greece, through the Roman Empire, the European empires between the 16th and the 19th centuries, to the hegemony of the United States in the second part of the 20th century, the control of knowledge has been one of the attributes in the system of power accumulation.

Many approaches that address the *power–knowledge* binomial as a core element of analysis have arisen from the social sciences. For the postmodernism approach, knowledge is shaped by the realities of power and, then, it also shapes them (Der Derian and Shapiro, 1989), which means that the actors that produce knowledge operate within a particular historical and political framework, in such a way that the categories they use are influenced by this same context. Based on Derrida's (1974) "deconstruction" strategy and Foucault's (1972) idea of a "truth regime," postmodernists try to analyze the factors that have allowed the creation of certain representations of reality (knowledge) and that have acquired a dominating position to the detriment

---

[1] Precedents of the link between knowledge and power can be traced back to Plato's theories on the "philosopher-king"; to Francis Bacon's *Meditationes Sacrae* (1597), where he stated that "scientia potentia est" (knowledge is power); and to Thomas Hobbes, who was quoted in 1658 saying: "the goal of knowledge is power […] the horizon of all thought is the undertaking of an action or thing to be done"; closer to our time, this topic has been of special interest to many theorists such as Gramsci, Marcuse, Habermas, or Foucault.

of others (power). Critical theory coincides with postmodernism on the fact that the production of scientific knowledge is always influenced by the social and political context. As Robert Cox (1993) points out: "Theory is always for someone, and for some purpose." Theorists of the critical approach use Max Horkheimer's concepts when they argue the existence of a "traditional theory," a follower of power, and a "critical theory," autonomous from power. Lastly, the feminist perspective adds to the debate the idea that the creation of knowledge is directly influenced by attributes that are considered as masculine, rationality, autonomy, and objectivity, which means creating explanations and knowledge that not only are reductionist and monocausal but also exclude feminine values (Youngs, 2004).

In the last decades, the relevance of scientific knowledge as an instrument of power started to grow. Alvin Toffler was one of the first to predict a radical change in the international system, as the world is moving towards a historic period in which human relations will be determined by the *power of knowledge*.[2] In the same way that, in the past, power shifted from weapons to money, scientific knowledge is currently replacing money. Toffler (1991) points out that the distance between advanced and less-advanced countries will be based on information and knowledge, and it prophesizes that future wars will be conflicts linked to knowledge.

The concept of *soft power* (Nye, 2010, 2017), understood as an alternative power resource to the classical *hard power* has emerged since the end of the '80s. This *soft power*, which includes knowledge as a substantial part of cultural, ideological, and technological media necessary for convincing and obtaining results, is now understood as superior to traditional military power. Susan Strange (1996) understands scientific knowledge as one of the basic pillars of the attainment of *structural power*, which gives power and authority to those people and institutions that occupy key positions in decision making to establish which knowledge is considered to be correct and desirable, as well as how it is used to generate consensus on the definition of problems and feasible solutions in an open context of uncertainty.

Currently, it is mostly understood that scientific knowledge has become a new instrument of power that has more and more influence on the dynamics of the international system and, as a consequence, the first discrepancies on how it is used have started to emerge. For some (Albornoz, 2001; UNESCO, 2005, 2015; Núñez Jover, 2006; Núñez Jover et al., 2020; Arocena, 2007, 2018; de Sousa Santos 2005, 2015, 2018; de Sousa Santos and Meneses, 2019),

---

[2] In his first book, 1970's *Future Shock*, Alvin Toffler was already expecting "many profound changes in a short period of time and the existence of an informational overload" arguments that he expanded on afterwards in 1980's *The Third Wave* and 1990's *Powershift*.

scientific knowledge is tightly linked with politics, because it has become a source and an instrument of extraordinary power and, as such, of interest for politics and interest groups.

> It seems indisputable that the growing role of knowledge is modeling a new structure of power, both on a global geopolitical scale and the social scale of each country or region [...] The links of domination of one nation over another, or of a social group over others, are increasingly related with the access to knowledge and the control on its generation and use. (Arocena, 2007)

On the contrary, other specialists such as Innerarity (2011, 2013) do not believe that knowledge, as an instrument of power, would be used to widen the social differences, but rather consider that STI and their diffusion in all aspects of daily life are allowing for the real possibility of defending against the powerful and of organizing opposition or learning to evade them, as, ultimately, knowledge increases everyone's capacity for action, not just the powerful. "We have grown accustomed to considering knowledge as an instrument to consolidate existing power relations as if the progress of science would always play in favor of the powerful, could be easily monopolized by them and could successfully remove all traditional forms of knowledge. I think that this idea of science as a repressive instrument that favors the powerful is inexact."

7.3 Mechanism of Social Innovation

Thirdly, we can point out a novel and increasingly relevant function carried out by scientific knowledge in the new international context, as a *mechanism of social innovation*. Since the end of the Cold War, the global agenda has become more extensive, complex, and interdependent, and requires a joint and sustained approach over time by all international actors to solve the great number of challenges that appear on a planetary scale. The issue, then, is the use of knowledge as a generator of creative ideas that allow the solving of problems in the extensive global agenda.

In recent decades, the evolution and development of scientific knowledge has extended the idea that this knowledge can be used as a mechanism to tackle humanity's common challenges. Knowledge, scientific research, technological development, and innovation have become the determining elements for economic development, improving health, taking care of the environment, and dealing with social problems of all kinds, such as poverty, social exclusion, or violence. Both governments and international organizations and institutions show a favorable opinion on the use of new advancements and innovations in scientific knowledge to solve global problems and, for this reason, most of the countries (both developed and developing) push forward policies that strengthen their knowledge production sectors and transfer and connect them to social and economic demands (Núñez Jover, 2006; Meek, Teichler, and Kearney, 2009).

From multiple political, economic, social, and academic fields, the need arises for scientific knowledge to be more extensively used as a tool for solving the countless *global problems* that humanity faces in the current historical moment, which transcend national frontiers and represent a significant threat for the international system as a whole. In this sense, scientific knowledge is understood as an increasingly useful and effective instrument in the search for specific solutions to the global challenges that the global agenda faces nowadays (environmental, energy, poverty, inequality, people movement, public health, etc.) through the creative application of wisdom and scientific knowledge for the solving of problems such as the production of food, machines, medicines, natural disasters or catastrophes, among many others. As the Royal Society points out, "Global challenges such as climate change, food, water, and energy security all feature highly on the agenda and require politicians to engage with science globally and locally in order to identify sustainable solutions" (The Royal Society, 2011). In the same line, Kuhlmann and Rip (2018) point out "Grand (Societal) Challenges, as a key part of the next generation of STI policy."

This innovative use of scientific knowledge understood as a social innovation mechanism to face the main challenges in the global agenda has had a fast evolution and strong support in the international system.

- In 1999, the first *World Conference on Science* in Budapest was organized by UNESCO, where a special reference was made to the prominence gained by scientific knowledge and the how as a tool to solve global problems. In its final declaration, it was stated that "the sciences should be at the service of humanity as a whole, and should contribute to providing everyone with a deeper understanding of nature and society, a better quality of life and a sustainable and healthy environment for present and future generations" (WSF, 1999).
- In September 2000, a large part of the most important and elemental needs of humanity were fixed by the United Nations in the so-called Millennium Goals, later revised and updated in 2015 with the *Sustainable Development Goals*. In both cases, the United Nations has asked its members to develop national policies aimed at promoting the production of STI applied to the global agenda, to achieve successful integration into the

knowledge economy through the use of new technology and sustainable development.
- The European Union has also manifested its concerns on the great global challenges and its open dedication to face them through the use of scientific knowledge. The *Lund Declaration* (2009) states that: "The global community is facing Grand Challenges. The European Knowledge Society must tackle these through the best analysis, powerful actions, and increased resources. Challenges must turn into sustainable solutions in areas such as global warming, tightening supplies of energy, water, and food, ageing societies, public health, pandemics, and security" (Lund Declaration, 2009).

Indeed, many of the international actors consider that scientific knowledge must be understood as a new global problem-solving mechanism and that, in the immediate future, it must deepen its role linked to social innovation. However, despite the great quantity of support and sympathy this idea generates, even declarations of intent are greater than concrete actions. The future scenery of ISR shows more complexity and multiplicity of conflicting interests, which is a serious obstacle when thinking of STI as a true and effective tool against global problems.

The great question that arises is, then, knowing if it would be possible to reach a level of STI governance that is sufficient to face the great magnitude of the global challenges. The following are some of those governability challenges:

- The multiple particular interests (often conflicting) of the many actors that intervene in ISR must be coordinated.
- A broad consensus must be reached to create scientific knowledge specifically focused on the international global agenda's most urgent topics.
- A global compromise must be established through juridical and administrative mechanisms that regulate and guarantee the application of STI to global challenges.
- The creation of a global and multilevel governance that guarantees the participation of all actors (both States and non-State).
- Mechanisms of actions that are accessible and efficient to assure the successful application of STI to human problems.

In short, STI can play a revolutionary role in the current international system as a tool for social innovation, as the fast evolution and development of modern science is allowing to reach concrete solutions to some of humanity's historical problems. However, there is still a long road to follow to consider that scientific knowledge is being effectively applied to the resolution of global problems. The main challenge is still the possibility of coordinating the interests of the multiple actors in the international system to transform humanity's problems and challenges into opportunities and benefits for society as a whole.

7.4 Democratizing Element

The fourth role scientific knowledge plays in the field of ISR is what is considered to be a *democratizing element*. In this sense, it is necessary to think about it more specifically as an enabler and generator of democratic uses, where STI can be used to stimulate democratic behaviors in the public space shared by a society (Arocena, 2007, 2018). This new function emerges in the context of a new knowledge society, which has allowed to increase individuals' empowerment, who now exercise their capacity to act in the field of democratic institutions that are more open to participation.

In the current global system, some positive uses that scientific knowledge has started to have as a democratizing element in society can be highlighted as follows:

(i) To inform, communicate, and spread the democratic system as a desirable form of government, as opposed to totalitarian regimes.
(ii) To help citizens discover, discern, and understand the orientation of the different political ideas offered with the goal of being able to choose based on common good or general interest.
(iii) To collaborate in the creation of democratic citizenship and political competences as a means to defend against abuses of power.
(iv) To boost different more participative styles of democracy, in which the methods of interaction between the different protagonists have real influence.
(v) To consider that knowledge opens a truly public space, that is, a space for meeting, debating, and democratic deliberation.
(vi) To boost the new possibilities offered by *e-democracy*, *e-administration*, and *e-government* through the virtual world and digital technologies to stimulate and ease the citizens' participation in public affairs.
(vii) To promote the usage of new technologies in the government to ease and improve the tasks carried out by the governments on different levels of action.
(viii) To stimulate decision-making processes in public policies based on scientific knowledge.
(ix) To understand the usage of the Internet and social media can potentially become a democratizing resource, stimulating the diffusion

of democratic ideas and practices through the participation and horizontal debate in the virtual world.

At the same time that scientific knowledge starts to play a more active role in ISR as a driving force behind democratic behaviors that help extend and deepen democratic participation, there are also risks and dangers for the bad use that can be made of these new scientific and technological developments:

- One of the main threats to the democratic use of knowledge comes from the economic spheres, which often seems to be attracting the most attention to the new role of scientific knowledge. Mario Albornoz (1997, 2007) clearly explains these dangers when he points out that "the culture of usefulness and benefit that prevails in the current globalization delays the true political debate on science and technology, substituting it with the attention to management, conceived from a strictly economic perspective. Market reason becomes the new Leviathan of the social order. The destiny of society is made subordinate to the demands of the possible clients."
- A second risk is related to the real capacity to extend the democratic use of scientific knowledge to the entire community. Historically, many societies have had an elitist concept of knowledge and have tried to exclude and reserve knowledge to restricted circles of privileged people. The relevance acquired by scientific knowledge in the new international context has further increased its value and threatens to transform it into an inaccessible resource for the majority of people. As UNESCO (2005) points out, knowledge can only become a true diffuser of democracy if "it guarantees universal access to knowledge, as well as the involvement of everyone in knowledge societies."
- Another critical point is the potential negative consequences that the installation of an electronic democracy could have. Some trends (cyber-pessimism and cyber-skepticism) have a negative vision of the phenomenon when considering that the Internet cannot become a democratizing factor because the distribution of scientific knowledge is increasingly unequal and that it only reproduces preexisting social inequalities.
- Finally, mistrust arises of the possible manipulation that citizens can be subject to through the use of new technologies, the Internet, and social media, which threaten to distort their democratic systems and their will when choosing their candidates. The suspicions of foreign insertion in the last two United States presidential elections (2016 and 2020) or the British Brexit referendum in 2016, with the deliberate goal of benefiting or harming one of the positions and manipulate the vote, clearly show the new dangers faced by democracy.

Whatever the case, what is evident is that scientific knowledge is increasingly used as a new democratizing element with a great potential to collaborate in the spread of democratic behaviors. The evolution and extension of this role in the international system will be decisive for reaching a full and global democracy.

7.5 Strategic-Military Factor

Finally, it is possible to identify a fifth role that scientific knowledge plays in the international system as a key factor in strategic-military development. What stems from the analysis of ISR at the beginning of the 21st century is the appearance of new ways of using innovative technologies in the military sector. Although there is a long tradition of applying for scientific and technological advances in this field, in the current international context it has broadened and intensified, and it has become a decisive factor for the international actors' strategic development of defense and security.

The usage of scientific and technological advances in the military field have been a constant in the history of international relations and the actors, almost without exception (be they empires, city-states, or Nation-States), have used technological advances as an essential input for their war machines to generate supremacy against rival armies and impose their strategic interests. Traditionally, the main hegemonic powers of the international system have been those actors that have known how to best make use of the technological advances applied to the military field.

In much the same way that the discovery of gunpowder or nuclear energy and their application to warfare caused a radical change in military strategy, with a great impact on the balance of power of the international system, the changes that are currently taking place in the field of ISR with the emergence of new disruptive technologies will presumably modify the international order once again. Based on this historical logic, nowadays all international actors are strongly investing in STI advances that can be applied in the military field, seeking to gain advantages that would assure them a better geopolitical positioning on the international system of the 21st century.

As has happened throughout history, advances in STI, in their most conventional shape (armies, weaponry, etc.), have a direct impact in the military sector where innovations are immediately applied to achieve tactical superiority over their rivals. Currently, technological advances in the military field have two main sources:

(i) The majority of the States' defense sectors have abundant resources to develop their own applied research which, in the case of some countries such as the United States and Russia, means almost 50% of total R&D investment.

(ii) Civil scientific and technological developments may also have their adaptation to the military sector. In fact, in the last few years, the government of the United States has recognized that, for the first time, its private sector is more advanced than the public sector in terms of scientific and technological advances and, for that reason, there is the need for a wider cooperation between sectors.[3]

The scientific and technological revolution in the last few years has substantially changed military tactics and strategy. Conventional combat has decreased notably, due to, in great part, mass communication media closely showing the horrors of war, which has generated a strong rejection in public opinion. At the same time, technological advances allow us to carry out a distance and remote-control war owing to the development of military drones, which allows the avoidance of troop and weaponry mobilization, thus considerably reducing soldier discharge.

The development of STI is revolutionizing the military industry by generating significant changes in all of its main aspects. Military armament is being substantially modified, which allows militaries to increase the capacity of destruction, to enhance precision, to increase automation, and even to strengthen the robotization of a great part of new equipment. Among the most outstanding advances are (i) autonomous and weaponized robotic vehicles; (ii) invisible and autonomous planes; (iii) drones used in the military field as true flying robots; (iv) remote-controlled autonomous weapons, strengthened by the development of artificial intelligence; (v) lighter, more mobile, and precise intelligent materials used in soldiers' equipment; and (vi) more lethal biological and biochemical weapons, owing to the advance in biotechnology and genetics.

The profile of armies in general and of soldiers in particular is also changing as a result of scientific and technological development. The traditional model of military recruitment, which forced the citizens of a country to form part of its national army, has lost importance against a professionalized model of military activity where armies will be smaller in number but better equipped. This phenomenon has promoted the rise of private military services that,

---

[3] Ashton Carter, United States Secretary of Defense (2015–17), recognized that the private sector's innovation capacities go beyond that of the military and that a more open relation with businesspeople, academia, and scientific institutions is vital to maintain the prominence of the American military capacities (Fojon, 2018).

under corporate logic, started to offer their services to the global market.[4] At the same time, soldiers have also benefited from new advances, which allow them to add to their equipment cutting-edge technology, including a system of geolocation and electronic sensors, costumes that will protect them from projectiles and other aggressions, and even provide them with superhuman strength, targeting elements that improve their ability to reach the enemy even when undercover; everything would be integrated into a rational operating system[5] (Cervera, 2016).

All these innovations applied to the fields of defense and security are promoting a readjustment of the geopolitical power balance on the international system. The United States maintained their military leadership, which represented an average of 35% of the world expenditure, triple that of China, the second world investor in defense, and seven times that of Russia, the third one (SIPRI, 2018). It is estimated that the total American defense expenditure, of approximately 649 billion dollars, is larger than the combined military budget of the eight nations that follow it in war spending. Currently, the United States has a nuclear arsenal 31 times larger than the United Kingdom's and 26 times larger than China's. China, the second-largest spender in the world, increased its military expenditure by 5% to $250 billion in 2018. This was the 24th consecutive year of increase in Chinese military expenditure (SIPRI, 2018).

However, in recent years, the countless advances in STI applied to the military field are increasing the doubts about the point to which the United States will be able to maintain its global military leadership. The emergence of technologies previously unthinkable, with a concrete application to the field of defense and security, such as artificial intelligence, machine learning, and the exploitation and analysis of big data, among others, opens a great question about which actors will make the best use of these technologies to improve their military system and also their geopolitical ranking in the international order. It is not a coincidence that, in recent years, countries such as China, Russia, Saudi Arabia, Brazil, or even the European Union have increased their defense R&D investment, understanding that future military supremacy will depend on the advances of STI.

---

[4] In the first decade of the 21st century, the private army market soared to become a business moving more than 100 billion dollars.
[5] Currently, there are multiple programs in the world that work with the development of technological innovations applied to soldiers' uniforms, including helmets with visors, special vests, new materials for uniforms, and integrated weaponry. Among the most relevant programs are the Future Force Warrior (United States), FIST (Great Britain), IDZ (Germany), Félin (France), Ratnik (Russia), and COMFUT (Spain) (Cervera, 2016).

In the United States, there is a deep awareness of this reality, and they have already started to act consequently.[6] In the so-called Third Offset Strategy, presented in 2014 by the then-Secretary of Defense, Chuck Hagel, the progress carried out by some State actors (China and Russia are specifically mentioned) and also some non-State actors in the development of military capacities is directly pointed out, and a long-term strategy to restore the global military power of the United States was proposed, through the production and acquisition of advanced weaponry systems and the development of new technologies applied to military industry (Martinage, 2014).

---

[6] In May 2015, the then-Chief of Staff of the US Air Force, Mark Welsh, admitted that, during the next five years, the military capacities of China and Russia will preponderantly increase, and that, in the future, China and Russia's military power could be even greater than that of the United States.

Chapter 8

CONFIGURATION OF INTERNATIONAL SCIENTIFIC RELATIONS

Having examined the different elements that form the ISR (Chapters 2 to 5), the new realities that have emerged within ISR (Chapter 6) and having defined the main roles adopted by scientific knowledge in the current international system (Chapter 7), the goal of this chapter is to explore the final configuration that ISR has acquired in the current world order.

A complete explanation of the current configuration of ISR requires a detailed description of two main elements of any world order:

- the *systemic parameters*, and
- the *global structure* that forms.

8.1 Systemic Parameters

Systemic parameters are all the constituent parts and elements of the system, which operate within the system's limits and, due to their dynamic nature, influence and are influenced by the system as a whole. All these parameters perform a key role in determining the final configuration of the system, and for this, it is necessary to analyze what kind of traits they transfer to the global configuration.

Actors

The first element of analysis in the current configuration of ISR is the international actors that have emerged as essential units that act and pursue their interests within the international system. In this sense, it is possible to define various characteristics of the international actors that influence the final configuration of ISR:

- A larger number of international actors show interest in STI (seen as a strategic resource for economic, political, and social development), which

is manifested in higher participation of actors in all processes related to scientific knowledge (production, intermediation, distribution, application, and governance).
- A larger number of international actors (both States and non-State) are interested in STI, which means an expansion of participants and interests.
- A larger diversity of actors taking part in the processes linked to scientific knowledge, be they States (including emerging and less developed countries) or non-State actors (such as transnational companies, epistemic communities, or think tanks).
- The university, the actor traditionally charged with producing and transmitting knowledge, must now share its tasks with other actors and take on new functions (the third mission), because of which it has started to develop new strategies to adapt to this new global context.
- The Nation-State still has a leading role in the creation of public STI policies, as well as in the planning and execution of strategic actions that stimulate its development.
- Sub-state entities are increasingly important, as they have reached a certain degree of autonomy and competency to carry out their own STI plans linked to the development of their local sphere.
- Intergovernmental organizations (IGOs) have been established as new prominent actors in the study, consultancy, organization, and execution of policies related to STI on a global level.
- Nongovernmental organizations (NGOs) have notably flourished in recent decades as agents with a special interest in the promotion and usage of STI as an element of problem solving for the international global agenda.
- Transnational companies have renewed their interest in knowledge and innovation as a main resource for the generation of economic benefits. This has led them to deploy active strategies not only to appropriate knowledge (property rights and patents) but also to produce it, either by associating with other actors (Triple Helix model) or on their own (academic capitalism and corporate universities).
- New STI non-State actors (such as think tanks, epistemic communities, and knowledge diasporas) emerge, which have become popular as experts and/or specialists in STI and that have become an authoritative source regarding the production, intermediation, and usage of knowledge.
- Finally, a *pluricentric* and/or *multicentric* context of knowledge has formed within ISR where more and more international actors (States and non-State actors) are showing interest in STI.

Interactions and relations

A second relevant aspect of the current configuration of ISR is the innovative and varied relations established among international actors. We can highlight the following global characteristics in the interactions and interrelations observed:

- The quantity and the intensity of all forms of linkage and relation (cooperation, conflict, competition, hegemony, and subordination) increase among STI actors.
- The conflictive links among actors increase, especially those related to appropriation rights, patents, and the privatization of scientific knowledge, as well as the tendency for attracting highly skilled personnel (talent).
- The competitive relations between international actors (States, companies, and universities) increase due to, in great part, the renewed economic value that scientific knowledge has acquired, which has caused a strong race among the main actors.
- Cooperative linkages increase thanks to the new developments in information, communication, and transport and the collaboration between multiple actors; also the emergence of new cooperation spaces (STI diplomacy, interuniversity cooperation, scientific diasporas, and Triple Helix) and the necessity of working together to tackle global challenges (social innovation).
- The relations of hegemony and subordination between different international actors, which established links marked by asymmetry and inequality, also increase. The so-called Big Three (the United States, the European Union, and China) and the Big Five (including Japan and Russia) still dominate the international system in STI and maintain the *cognitive divide* and *dependency*.
- New linking mechanisms among actors appear (cooperation networks), and some traditional linkages are renewed and strengthened (STI diplomacy).

Processes

A third element that shapes the current configuration of ISR is the deep changes that happened in the internal processes and mechanisms. In each of these scientific knowledge processes (production, intermediation, distribution, application, and governance), new particularities have been generated and established:

Production

- The methods of STI production have evolved and strongly changed in recent years, generating new ways of creating scientific knowledge (innovation systems; Triple, Quadruple, and Quintuple Helix, Mode 2; Mode 3, postnormal science; social development).
- The methods of scientific knowledge production now include in their process a larger number of actors and participants (especially non-States), a larger quantity of interaction between actors (two- and three-way interactions), and new places for knowledge creation (not only the university).
- Knowledge creation is now done in a more interdisciplinary way (new sciences and subdisciplines), application-oriented (applied research), and with different quality controls (not only scientific peer review).
- New fields of application and purposes for what is produced (especially oriented to the economic sector).
- A great increase in all the production factors of scientific knowledge (R&D investments, researchers, highly skilled workers, higher education investment, research infrastructure development, publications, and patents).

Intermediation

- An increase in the quantity, speed, and intensity of processes of scientific knowledge intermediation.
- The mediation of scientific knowledge is now carried out in different ways to accomplish the many objectives asked of it (transference, transmission, diffusion, or mobility).
- Intermediation has three main goals: (i) training skilled personnel, (ii) transferring to productive sectors, and (iii) diffusing to the entire society.
- A larger participation of actors (not only Nation-States but also non-State actors), which now fulfill outstanding functions throughout the intermediation process.
- Scientific knowledge is transferred to society through institutions and formal education (which implies new challenges for higher education), new ways of communicating scientific research, new processes of teaching and life-long and online learning, and new ways of transmitting information through cooperation networks.
- The diffusion of scientific knowledge has spread thanks to the new possibilities offered by new digital information and communication

technologies, which have led to the rise of cooperative networks and the virtual world.
- Rise in the transfer of knowledge to productive sectors for its transformation into innovations that generate economic benefit and development.
- The encouragement of the mobility of highly skilled personnel throughout the international system grows as a more effective way of transporting tacit knowledge.

Distribution

- The increase in the number of international actors that participate in the processes of knowledge, whether States (emerging countries) or non-States (mainly nontraditional ones such as think tanks or epistemic communities).
- The processes of scientific knowledge distribution show a slight expansion of the distribution of knowledge to new geographical areas, which implies the emergence of new countries and regions interested in STI.
- The distribution of STI is heavily unequal and concentrated in certain centers and/or nodes on all levels of analysis (State, regional, intraregional, and economic development).
- The distribution of the main factors of scientific knowledge production (R&D investments, number of researchers, scientific publications, patents, academic institutions, and destinations of international students) is concentrated in a few geopolitical nodes/hubs.

Application

- Application has especially focused on socioeconomic objectives related to the commercialization of STI.
- There is a strong trend towards applied research and experimental development over basic research.
- The application of scientific knowledge in specific scientific areas such as engineering, technology, and health sciences, and the new areas of science and technology such as nanotechnology, robotics, and artificial intelligence.
- Increase in the number of scientific publications and patent applications in STI fields, especially engineering, biological, and medical sciences, information technology, and digital technology and communication.
- The investment and applications in the defense and security and military sectors are high and constant.
- The private sector leads the application of emerging technologies in many countries (including the United States, Canada, and Australia), with a

strong trend towards investing in the software and Internet, health and pharmaceuticals, vehicles, and information and electronics sectors.

Governance

- The organization and planning of scientific knowledge have gone from types of government and management associated with only a few actors (mainly States and companies) to a true process of global governance.
- A larger participation in the debate and articulation of public policies of multiple actors (cities, states, companies, regional processes, IGOs, NGOs, etc.) that are now interested in their organization and coordination.
- States still have an important role in the governance of scientific knowledge as catalysts and articulators of the multiple conflicting interests that international actors have on the organization, execution, and evaluation of public policies.
- In recent decades, the majority of countries have started to develop ambitious public programs to promote STI; some of them have started to design new programs aimed at investing in emerging technologies applied to economic and social development (Germany, China, India, etc.).
- Cities and regions emerge as local actors interested in the planning and execution of public projects linked to STI (Quebec, Basque Country, etc.)
- Integration processes have developed new methods of regional governance by establishing cooperative agreements among their members for the creation of common scientific policies (the European Union)
- In recent decades, governance at the international level has strongly developed through a multiplicity and variety of linking processes (continental, trans-sovereign, transnational).

Emerging realities

New realities have emerged in the ISR and allow us to identify some of the main particularities that define the current global configuration of ISR. The following are among these new systemic realities:

- The existence of knowledge divides (digital, cognitive, and scientific), which are the product of the unequal distribution of STI on a geopolitical and geoeconomic level, creating new knowledge centers and peripheries.
- The existence of a governance that is global (as it implicates practically all actors) and multilevel (as it is carried out simultaneously and interrelatedly in various fields and arenas).

- A significant increase in the commercialization of STI due to the economic revaluation of scientific knowledge and the intensification of the competition between the different actors of the international system, which has generated an expansion of the market for all goods, services, and processes derived from STI.
- An improvement in the participation of women in STI, although there is still an evident gender divide to overcome.
- The emergence of new areas in science and technology, which are generating a true scientific revolution, which means deep disciplinary changes, new technical applications, and a strong economic, social, and political impact.
- A strong but constant geopolitical and geoeconomic shift in the field of STI is modifying the traditional Western supremacy with the rise of East Asian countries (especially China).
- The emergence of a new virtual world towards which a large part of human activities is moving, which represents great opportunities but, at the same time, significant threats to the security of the entire international system.

8.2 Global Structure

A second element needed to analyze the global configuration of ISR is the global structure the system adopts. This structure is manifested as a result of the interaction between actors, which ends up shaping a power system or global order that characterizes the system as a whole and which normally determines dynamic domination and subordination relations among actors. In this sense, it is very important to know how this process of power rebalancing takes place among the main actors in ISR and which type of global structure they built.

Polarization

The process of polarization between the main actors related to STI is one of the most important traits of the new configuration of ISR. Essentially, polarization is a very dynamic process by which the main actors of the international system fight and compete between them trying to establish power balances in the system.[1]

---

[1] Not all members of the international system take part in a polarization processes; it is a phenomenon mainly established by those with the capacity of creating the rules of the game or, at least, of modifying them (Dallanegra Pedraza, 1998).

The study carried out on the behavior of actors in ISR allows us to see a triple dimension with regards to the process of polarization: (i) the polarization between Nation-States, (ii) the balance between States and the rest of non-State actors, and (iii) the polarization between non-State actors.

(i) The process of *polarization between States* shows notable changes in recent years. These modifications can be summarized with the following particularities:

— The United States maintains world leadership in STI, although its hegemony is now seriously threatened by China's exponential growth and the European Union's advances.
— The triad or Big Three (the United States, the European Union, and China) and the Big Five (the Big Three together with Japan and Russia) still have a privileged position with regards to other countries.
— There has been extraordinary growth in one State in particular (China) that, in a few years, has placed itself at the forefront of STI development.
— New intermediate countries and/or powers (Brazil, India, South Korea, etc.) have emerged, which make a strong commitment to knowledge and to reduce the differences with developed countries.
— The distance increases between the most developed and least developed countries, which still have a very unfavorable global position.
— The polarization in the field of ISR shows a global structure tending towards a system with increasingly multipolar characteristics. It is a transition from the hegemony of the United States and the traditional triad towards a more extensive and distributed configuration where more States (Brazil, India, Russia, South Korea, or South Africa) are taking part but that, at the same time, is becoming more unequal when considering all actors (between developed and poor countries).

(ii) The process of *power balance between States and the rest of non-State actors* also shows new characteristics that affect the global configuration of ISR:
— States lose their exclusive monopoly and control over the majority of the STI processes (which they have enjoyed historically), now held by new non-State actors (transnational companies, IGOs, etc.).
— New non-State actors interested in scientific knowledge emerge, which are interested in occupying leading roles in all of the STI processes (NGOs, IGOs, companies, epistemic communities, think tanks, and scientific diasporas).

- Conflict, cooperation, and competition links between State and non-State actors are rapidly growing due to the diverse interests they follow (they cooperate and compete, at the same time, for new emerging technologies).
- Some transnational companies have reached capacity and power similar to many of the more developed countries, which makes the polarization between them more intense and competitive.

(iii) Lastly, there is a new process of *polarization between non-State actors* due to a strong competitive race in which most of them participate.

- Universities show a polarization similar to that of States, with a high concentration in the so-called world-class universities, which are distributed around a few countries (the United States, the United Kingdom, and Japan).
- Think tanks are also polarized as they have to compete between themselves for funding sources, economic aid, political links, and access to markets for their services.
- Companies (which naturally compete for markets and benefits) are also showing notable polarizations, mainly manifesting in the capacity they develop to produce scientific knowledge and turn it into innovations that can be applied to the market.
- Individuals linked to STI, whether researchers, professors, or students, are also becoming polarized for resources, offers, and opportunities.
- Finally, there is cross-polarization among different actors; for example, in the process of public policy design, where everyone interacts and competes seeking to impose their own interests.

In the three cases of polarization (between States, between State and non-State actors, and between non-State actors), the current configuration of ISR shows a clear trend towards *multipolarization* with the rise of new both State and non-State actors that take an active part in all processes related to STI and assume central positions in the global configuration.

Distribution

Another point of interest for the current global configuration of ISR is the geopolitical and economic distribution of scientific knowledge in the international system. Among this distribution's main characteristics, we find the following:

- Although many actors are starting to join this new global configuration of ISR, increasing the participation of State and non-State actors, the global structure of ISR is still very unequal in terms of STI distribution.
- The inequalities happen on different levels: between countries (the United States, China, or the European Union with the rest of the world), between regions (East Asia, North America, and the European Union with others), between regions on the same continent (in Africa, America, or Europe), and is also observable within the same country (e.g., Buenos Aires in Argentina, Moscow in Russia, or San Francisco in the United States).
- The differences in the geographical distribution of the main educational centers and universities (the United States, the United Kingdom, and Japan), think tanks (the United States and Europe), or companies (North America, Europe, and Southeast Asia) are also increasing and expanding.
- The number of researchers, publications and patents, R&D investments, world-class universities, and international student destinations shows a distribution highly concentrated towards more developed countries (led by the United States).
- The inequality in the distribution is also observed in the mobility of highly skilled personnel or *talent*, whose flow is mainly focused towards the most developed countries due to the better life and work conditions they offer and the active attraction policies implemented by these countries (the European Union, Canada, or the United Kingdom).

In short, we can see the existence of an *unequal distribution* on a global scale that is shaping a new global configuration of ISR that is characterized by the creation of *new divides* (cognitive and scientific) and the expansion of the processes of *centrality* and *marginality of scientific knowledge*.

Hierarchization

As a result of the polarization and unequal distribution of scientific knowledge, a clear hierarchization between actors has been established as another characteristic of this new global structure of ISR. This means a classification and/or categorization by levels of importance based on the access to scientific knowledge, which is visible in the field of STI:

- States maintain a clear hierarchization between them that shows the United States, China, some European countries (mainly Germany, the United Kingdom, and France) and Japan at the top of this hierarchy;

emerging regional powers such as South Korea, Brazil, India, and South Africa in second place; and, lastly, an extensive group of less developed countries that also show differences among them.
- Regions also have this hierarchization: North America (the United States and Canada), the European Union and East Asia (mainly China and Japan) are the most relevant zones; in second place, some Latin American and Eastern European countries and new emerging Southeast Asian countries; lastly, the less favored regions in Africa, Asia, and Latin America.
- Companies equally show a hierarchization based on their size, innovation capacity, and access to the international market.
- For their part, *universities* have formed hierarchies due to the strong competition between them, which has stimulated the creation of global rankings that compare their performance in education, teaching, and research, thus establishing a clear hierarchization among them (world-class universities).
- Think tanks and NGOs also have their own rankings that classify these actors according to their development, capacity, and efficiency in the consulting and advisory tasks they carry out (the Think Tanks and Civil Societies Program (TTCSP) and Global Geneva).
- Cities are also entering an international hierarchy according to their link with STI and the capacity they have for attracting actors, relations, and innovating processes to their geographic spaces (innovation hubs, such as Silicon Valley, San Francisco).

Segmentation

The phenomenon of hierarchization implies, at the same time, a process of *segmentation* or division into sections, in which actors are differentiated according to the access to scientific knowledge. This process can be observed between

- States that segment according to scientific knowledge into developed, developing (or emerging), and poor countries;
- universities are segmented based on their ranking, forming differentiated groups such as "top 10," "the 100 best universities," or "the 500 best universities"; and
- think tanks and NGOs are also creating rankings that evaluate their performance and then differentiate them into performance and quality segments.

Sectorization

Finally, a topic that also characterizes the current global structure of ISR is the areas or fields where knowledge is produced, applied, and used it. In this sense, the main particularities it presents are the following:

- Increasing marketing, commercialization, and privatization of STI in the context of a knowledge economy, where knowledge is seen as a fundamental element for the generation of innovation for the attainment of economic benefits.
- The idea of scientific knowledge as a resource of global power is growing, which shows the novelty of being a soft power, although just as effective as the so-called *hard* power.
- The application and usage of scientific knowledge increases in the fields of applied research, experimental development, technical disciplines, and the so-called new areas in science and technology (such as artificial intelligence, nanotechnology, or biotechnology).
- Constant high levels of investment and expenditure in defense, security, and military sectors are maintained.
- The incipient use of scientific knowledge as a mechanism of social innovation and/or tool to tackle the challenges in the current global agenda.
- The usage of scientific knowledge as a factor to extend and deepen democratic processes on a global level, as well as a tool to improve the reach, efficiency, and productivity of public administration (e-government).

To conclude, the in-depth analysis of the ISR showed relevant changes in the *systemic parameters* (actors, relations, processes, and new realities) and also a new configuration of the *global structure* (more polarization, unequal distribution, hierarchization, segmentation, and sectorization).

Chapter 9

INTERNATIONAL SCIENTIFIC RELATIONS AND THE INTERNATIONAL SYSTEM

After the in-depth analysis of the new roles played by knowledge in the current international system (Chapter 7) and studying the main characteristics of the current global configuration of ISR (Chapter 8), the last thing to do is to identify the main interactions and the mutual impact established between ISR and the international system. The goal is to examine how ISR, understood as a specific field or subsystem in international relations, links with other subsystems and with the international system as a whole.

The use of the systemic methodology has allowed us up to this point to analyze ISR as a dynamic and complex system whose parts interact between themselves, develop internal processes, and conceive new emerging realities that end up characterizing the system as a whole. However, the same systemic perspective allows us to go further with the analysis through positioning ISR within a wider system (the international system) that contains it and conditions it and where, at the same time, it interacts with other subsystems such as the economic, political or social systems. This allows us to see how ISR links with the other subsystems that form its environment, as well as to discover how it interacts with the international system as a whole.

Within this conceptual and methodological framework, this last chapter seeks to answer three fundamental questions:

(i) How does ISR interact with other subsystems (political, social, economic, military)?
(ii) What are the most relevant trends generated by ISR and how do they feed back into the international system?
(iii) What impact does ISR have on the 21st-century World Order?

The main goal behind these questions is to identify the main demands that ISR receives from other subsystems and know their answers; to understand how these demands are processed within ISR and how they become trends that affect the international system as a whole; and, lastly, to recognize the impact ISR has on the world order that is forming at the beginning of the 21st century.

9.1 Interactions with other Subsystems

The first question to be answered concerns the interactions established between ISR and other subsystems in the world order. In the systemic vision, the subsystems of a system are intimately linked establishing frequent contacts between them through crossed demands and replies. This is how ISR, understood as a subsystem in the international system, receives specific demands from other subsystems, which it internally processes and to which it later gives a timely reply, feeding back into and impacting those systems and the international system as a whole. Within these systemic feedback processes, we highlight the interactions between ISR and the economic, political, strategic-military, and social subsystems.

Economic system

As has been previously stated, within the economic field there have been deep changes in recent decades, as a product of the transformations and evolution of the capitalist economic system, and the impact generated by the scientific and technological revolution. In this new phase of the capitalist system, scientific knowledge has acquired an essential role as a new production factor and a strategic resource for the creation of innovations that allow companies to compete on a global level. The fact that scientific knowledge has become an essential element for the economic system gives all linkages and processes generated within ISR significant effects in the economic field. In this context, the multiple demands made from the economic system on ISR can be summarized in three items:

– The generation of more abundant and complex scientific knowledge, to be used as an input in the production of innovative goods, services, and processes with commercial value.
– A larger articulation of interests and an intensification of the cooperation between actors in both subsystems for the development of scientific research geared towards the productive sector.
– The intensification and acceleration of the qualification and training of qualified personnel that may join the productive economic system as specialized workers.

ISR has answered these demands through a variety of actions and processes that have a highly positive impact on the economic subsystem. Among these answers, we highlight (i) a higher investment by the majority of actors in R&D in researchers and highly skilled workers; (ii) a higher quantity and diversity in the production of scientific knowledge; (iii) the emergence of new methods of knowledge production oriented towards the generation of innovations linked to the economic sector; (iv) rise of new intermediation mechanisms, such as scientific and technological transfers to the productive sectors; (v) a larger quantity of applications to disciplines, types of research and areas directly linked with economic development; (vi) larger participation of companies in the production and mediation of knowledge; (vii) a larger boost to academic capitalism; (viii) a larger development of emerging technologies that generate innovation; (ix) a larger quantity and diversity of educational programs and training for undergraduate and graduate students; and (x) a greater degree of collaboration and cooperation between companies and scientific sectors.

Political system

We have also analyzed in depth the major changes in the international political system. Many political processes have occurred in the last decades: the end of the Cold War order, the rise of new State and non-State actors, appearance of new levels or arenas of action, slow but constant expansion of democracy, and a strong reaction to the globalization process that has boosted strong phenomena of nationalism, populism, and/or terrorism throughout the planet (Rodrik, 2011, 2020). In this context, demands to ISR are constant and mainly linked to the usage of this new knowledge as a political tool:

- Knowledge and STI have started to be seen as a new and strategic factor for accumulating and maintaining *soft power*, which is now as valuable as the traditional hard power.
- New information and communications technology are asked to play a more active role in the promotion of democratic usages and customs in society.
- New digital technology, used in new devices and social media, is also seen as a new and effective way of reaching citizens to showcase and promote political views and influence their political-electoral behavior.
- At the same time, these technologies are being used to spread extremist ideas, and accordingly, there is a demand for new technological mechanisms to combat fundamentalists.
- Lastly, it is understood that STI can offer specific solutions to the large amount of government administrative and bureaucratic tasks to make

them more agile and efficient through the usage of new technologies and digital formats.

ISR replies to these demands through multiple actions with strong positive repercussions within the political system. The most important responses are that (i) scientific knowledge is increasingly becoming a new factor of *soft power* that is used by all actors that want to make the best use of this new resource to empower themselves and achieve their own objectives and interests; (ii) the development of new technological media (mainly the expansion of connectivity, the usage of the Internet, and the spread of digital devices), which are used to spread and democratize information and knowledge and, at the same time, are used as resources for the promotion of democratic processes and customs; (iii) new digital technologies and social network platforms that have become a new and powerful political tool that allows for the spread of political ideas, but that also for the use of user data and profiles to influence their opinions and electoral behavior; and, lastly, (iv) scientific advances that have allowed the development of new technological tools, which are now applied by the government to speed up official administration and enable citizens' participation in political affairs through new phenomena such as *e-administration* and *e-government*.[1]

Strategic-military system

Historically, the strategic-military system has demanded the latest technological advances from ISR to apply them to its defensive apparatus. The intensification of the scientific and technological revolution and the transition to a new post–Cold War international scene, where new forms of polarization have emerged among actors, has given new value to the role played by STI in the military sector. In this new context, the main demands of the strategic-military subsystem to ISR are focused on four key points:

– The changes in the nature of security and war itself demand the usage of new STI (especially new emerging technologies) as an indispensable tool for the development of military power.
– There is a demand to maintain and/or increase the R&D investment linked to the fields of defense and security to face the race for technological

---

[1] Since 1997, Estonia's government has developed the *e-Estonia* project, which started with the digitalization of administration (e-governance) and, over time, has extended to new sectors linking the government with the citizens: e-tax (taxes), e-voting (electronic vote), or digital ID (identity documents).

development and geopolitical changes that are taking place in the international system of the 21st century.
- There is a need for solutions to tackle the new challenge represented by the virtual world, from which it is now possible to attack and cause irreversible damages to a country or region.
- There is a demand for larger cooperation between the private and military sectors, intending to create synergies allowing for innovations reached by many companies to find a faster application in the military field.

Nowadays, the military sector's demands towards ISR are met positively through knowledge and innovation production, with a high degree of application and usefulness for the defense and security sectors. Among these responses, the following stand out: (i) the investment in the production of scientific knowledge related to the strategic defense, security, and the military sectors has been sustained or even increased, with a high expenditure (an average of 30% in OECD countries, and going up to 50% in countries such as the United States or Russia); (ii) the number of investments in new technologies such as big data and artificial intelligence applied to the military field with the goal of strengthening existing virtual defense systems is increasingly relevant; (iii) the development of STI applied to armaments, structures (both physical and virtual), and military personnel continues to be driven by the objective of perfecting material and human resources for physical combat; and, finally, (iv) private sectors are increasingly connecting with public sectors in the search of joint military developments.

Social system

The current social system has been deeply influenced and modified by scientific and technological development to the point of considering the current society as a *knowledge society*. In this context, the main demands from the social system are focused on making social use of the newly developed knowledge and improving the impact the adoption of STI has on social life:

- The social system demands scientific and technical solutions to tackle global challenges, seeing scientific knowledge as a possible innovating mechanism to successfully face the global agenda and achieve sustainable development for the entire international system.
- A more equal distribution of scientific knowledge is also demanded due to the strong geopolitical and geoeconomic inequalities shown by the current international system.

– The social impact of STI is also generating strong demands for new emerging technologies to spread its innovation not only to the military and economic sectors but also to society as a whole with concrete applications for basic tasks of daily life.

ISR replies to these demands in different ways: (i) developing and perfecting the production, intermediation, and application of scientific knowledge with the goal of giving new and more efficient solutions to the topics of the international agenda. In this sense, scientific knowledge is increasingly used by multiple actors as a new resource for searching for concrete solutions to the most relevant social issues (poverty, inequality, pollution, etc.); (ii) there are also many actors linked to knowledge (even inside the tech industry itself) that in recent years are advocating for guaranteeing a minimum degree of equality in STI distribution through a new paradigm that would combine STI development with distributive justice criteria such as the ones proposed by "Technological Justice" (Ortega and Perez, 2018) or the "Society 5.0" paradigm[2] (Keidanren, 2016; Fukuda, 2019); and, lastly, (iii) scientific and technological progress is increasingly concentrated in developments with concrete applications to people's lives, easing their daily tasks (technology applied to the home, autonomous vehicles, public transport systems, etc.).

9.2 Macrotrends

To respond to the second question (*What are the most relevant trends generated by ISR and how do they feed back to the international system?*), we must understand that the demands generated by the different subsystems and the international system as a whole have a strong impact on ISR that internally process these demands and turn them into responses and macrotrends that will feed back into the entire international system.

In the systemic perspective, feedback and self-regulation are intimately connected as the trends in a subsystem feedback to the whole of the system in the shape of *negative feedback* (continuity trends) or *positive feedback* (change trends). *Continuity trends* have a very important role in obtaining and maintaining stability in a system by perpetuating its *status quo*; for their part, *change trends* increase the possibility for innovation and, as such, they benefit change, affecting the loss of stability or balance of the system.

---

[2] The Japanese plan "Society 5.0" has as its main objective the digitization of not only the economy, but all levels of Japanese society, seeking the (digital) transformation of society itself (Fukuda, 2019).

According to the observations throughout the systemic analysis, many macrotrends are generated within ISR and that promotes both continuity and change in the international system's *status quo*. The following are the most important macrotrends:

Continuity macrotrends

– The reevaluation and empowerment of traditional actors (States, companies, and universities).
– The deepening of traditional conflictive links (property rights, patents, or talent attraction policies).
– Stimulus to the competitive economic system (race between actors for STI).
– Persistence of asymmetric links, which create relationships of hegemony and subordination between actors (inequalities and scientific dependency between countries, universities, or companies).
– Conservation of the gender divides in STI despite women's higher participation in higher education and the development of public policies aiming to break existing barriers (glass ceiling and sticky floor).
– Extension and increase of social inequalities (through digital, cognitive, and scientific divides).
– Strengthening of traditional methods of governance (at national, regional, and international levels).
– Promotion of higher education, universities, and international mobility of researchers and students (larger investment from public and private sectors).
– Deepening of the marketing and commercialization processes (larger application of scientific knowledge to the economic sector).
– Persistence in the fight among the main international actors for world talent (with public and private attraction plans).
– Contribution to the national defense and security sector (R&D military investments are stable).
– The Nation-State maintains an essential role in the generation, planning, and execution of public policies related to STI (national plans).

Change macrotrends

– The emergence of new actors (epistemic communities, scientific diasporas, etc.).
– Reevaluation and strengthening of secondary actors (IGOs, NGOs, and sub-State entities).

- New models of cooperative interaction (STI diplomacy; physical and virtual cooperation networks; Triple, Quadruple, and Quintuple Helix; innovation helix framework model).
- Stimulus in more complex governance processes (by areas or levels, formal and informal, with a multiplicity of actors and topics).
- Strengthening of innovative political power instruments (soft power).
- New ways of creating and producing scientific knowledge (new models that involve more participation, modalities, topics, places, etc.).
- The appearance of new economic production factors (scientific knowledge and innovation).
- New ways of knowledge intermediation through the transference to productive and commercial sectors (science marketing).
- Development of new scientific answers to face the global agenda (scientific knowledge used as an instrument for social innovation).
- Contribution to the expansion of democratic governments (using STI as a democratizing element).
- Development of new digital information and communication technologies (larger investment and application in this sector).
- The appearance of a virtual world or cyberspace creates a new field of action for all human and social activities (Internet, social media, and social networks).
- Significant geopolitical changes that show a shift of STI concentration from the West to East (China's leading role).
- Stimulus to new ways of governance (multilevel and global).
- Emergence of new areas in science and technology with a high economic, social, and political impact (artificial intelligence, robotics, nanotechnology, etc.).

In sum, ISR is influencing the international system as a whole through very concrete responses that become continuity or change macrotrends that operate by feeding back to the balance and regulation of the entire international system. All these replies imply the inclusion of larger elements of complexity in the world order due to the large quantity and variety of actors, relations, processes, phenomena, and realities that emerge from the field of ISR to the international system.

9.3 Implication for the International System

To answer the third and final question (*What impact does ISR have on the 21st-century world order?*), it will be necessary to recognize and describe how ISR and STI influence the international system as a whole. The goal is to discern how the trends generated in the field of ISR end up impacting and modifying the international system as a whole, which will allow us to better understand the consequences that everything happening within ISR have in the shaping of a new world order.

The systemic analysis carried out throughout this research has allowed us to understand that, as happens in every complex system such as the international system, processes are essentially multidimensional and multicausal, extraordinarily interrelated, and with notable consequences for all actors involved. For this reason, ISR receives multiple demands from the global context formed by the international system and from the many subsystems that form it (economic, political, strategic-military, and social), internally triggering a systemic transformation process of all those inputs into responses (macrotrends) that feed back to the entire international system.

The processes generated in the field of ISR become a set of macrotrends (continuity and change) that, by acting simultaneously, affect the regulation of the global system as a whole and, especially, in the way the world order is being configured. The main consequences can be summarized in two major types of impacts:

(i) *Modification of systemic parameters*: The macrotrends originating in the field of ISR especially affect each and every one of the parameters and elements that compose the international system. These macrotrends have a strong influence on

- *actors* (through the emergence of new actors and the empowerment of traditional actors);
- *relations* and *interactions* (increasing the quantity, intensity, and modality of the links);
- *global agenda's topics* (expanding the spectrum of relevant issues);
- *processes* (boosting new and complex mechanisms);
- *phenomena* (stimulating the emergence of new emerging realities).

(ii) *Changes in the global configuration of the international system:* In parallel, there is a second systemic impact which, in this case, directly affects the global configuration of the international system. These consequences can be observed in

- the modification of the *polarization process* (with a clear trend towards multipolarization through the empowerment of new State and non-State actors);
- the strengthening of the *inequality divides* (with the emergence of knowledge divides that stimulate the creation of centers and peripheries and social divisions as well as fortify old social gaps);

- the increase in *hierarchization* and *segmentation* between actors (with different levels of classification and categorization);
- the sharp *sectorization* of STI (with developments and applications in the economic, political, social, and military domains);
- the stimulus of some *scientific areas* and *fields* over others (prevalence of the productive sector over social relevance);
- the development of new *emerging technologies* with great disruptive power (artificial intelligence, robotics, etc.);
- the appearance of a new global *power resource* (science, technology, and innovation).

These two great impacts caused by ISR on the parameters and global configuration of the international system allow us to recognize one last fundamental consequence: *the existence of a new and privileged place for STI in the international system*. In this sense, having studied in detail the complex system that ISR represents, it is possible to conclude that scientific knowledge has acquired a new prominence in the new world order in at least two senses:

- *Science, technology, and innovation has become the main and most strategic resource for the international system*: In the current context of ISR, scientific knowledge has been revalued as a key input and a strategic resource for the international system, becoming the main source for generating economic benefit, political power, social innovation, and military development, which has allowed it to play a more important role than the one it has traditionally had. What is distinctive in this new historical period is that STI is found at the intersection of all international processes, strongly influencing sectors in economy, politics, and society and substantially modifying productive processes, the labor market, the way of doing politics, the way of spreading out and communicating, the nature of war, and practically all the basic aspects of daily life.
- *Science, technology, and innovation has also become the new ordering principle of the international system as a whole*: Although at other historical moments, religious, military or economic factors had leading roles as the guiding principles and main drivers behind the international system, currently, this key function has been taken over by scientific knowledge. ISR has become a key part of the international system and is turning STI into the core element of the main systemic interactions. In this way, scientific knowledge not only becomes a fundamental strategic resource but also

the main factor around which a large part of the relations and processes within the international system revolve.

In the current context of a transitioning international system, of which many characteristics are still unknown, it is at least possible to be sure that, in the 21st century's world orders, STI will have a decisive role and ISR will be the most relevant field in the international system.

CONCLUSIONS

The research carried out to this point has been presented to the reader as an ambitious intellectual work, but also being conscious of the difficulties and limitations faced when analyzing an issue so complex as the link between STI and international relations. Facing such a challenge, the author has carried out a research project approaching a subject of study (science, technology, and innovation), analyzed a subdisciplinary field (international scientific relations), used a disciplinary perspective (focused on international studies, but also openly interdisciplinary), used a methodological focus (systemic paradigm), and applied research techniques (systemic models) that have shown a great descriptive, analytic, and explicative potential. The challenge and the intellectual enterprise were huge, but the social and academic need for specific answers encourages the realization of this project.

The main objective of this research has been studying the empirical space emerging from the interaction between STI and international relations. The analysis has shown three important findings:

- first, the existence of new actors, relations, processes, and phenomena, and a strong agenda that demands more scientific treatment;
- second, the confirmation of a promising and vigorous field of study called *international scientific relations*;
- third, the fact that STI as a subject matter requires new and deeper approaches from the field of international studies.

Based on this new empirical and academic context, it is pertinent to introduce this research that tries to contribute in the same direction as those scientific works that are encouraged to face the current global challenges. The final result has been an extensive, detailed, and holistic analysis of the new reality of scientific knowledge within the international system and the main changes and continuities that are currently happening in the subfield of ISR. This has allowed us to offer answers to three main problems: (i) the role that STI plays in the current international system, (ii) the main characteristics shown by the current global configuration of ISR, and (iii) the impact that STI and ISR have on the 21st century's international system.

Role of science, technology, and innovation

STI is a determining factor in the 21st century's international system. Although STI has been a relevant and strategic element throughout a great part of the history of international relations, the second part of the 20th century has seen a great boost as a consequence of a combination of factors endogenous to the international system (*big data* policies, evolution of the capitalist system, the rise of globalization, acceleration of the scientific and technological revolution, etc.). From there on, STI's relevance has not stopped growing and, at the beginning of the 21st century, the world is rapidly changing, due in large part to the magnitude, intensity, and ramifications scientific knowledge has acquired as a key factor in economic, political, military and social development. The scientific and technological revolution has opened a new stage with the discovery, development, and application of new areas in science and technology that promise an extraordinary impact in all aspects of society. As a result, STI has acquired new roles and has assumed an indisputable leading role and influence in the global dynamics of international relations that has made it a decisive factor in the international system of the 21st century.

Throughout this research, it was possible to establish at least five primary roles and functions played by STI in the current international system. The reevaluation of scientific knowledge in recent decades has intensified some of its traditional roles, at the same time that it has stimulated new functions:

- *Resource of economic gain*: Although STI has been used in this way in the economy at least since the Industrial Revolution, what is new is its essential role as a resource for obtaining economic benefit in this new stage of the capitalist economic system. Currently, scientific and technological research and its product, scientific knowledge, is the main input for the generation of innovations (product, processes, and services), which, at the same time, are the key element for the new competitiveness between companies, countries, and regions in the context of a knowledge economy. This new reality has triggered the interest and investment of the most relevant actors in the international system (and not only companies) for the production and control of STI.
- *Instrument of political power*: The conception of knowledge as an element that gives power to whoever has it is as old as the 16th century quote attributed to Francis Bacon ("knowledge is power"). However, what is new in the current international stage is the confirmation of STI as

a critical factor in the international system to build and accrue power, thus replacing old power archetypes of power such as military, religious, and even economic power. Scientific knowledge extends as part of the *soft power* used by cultural, ideological, and scientific sectors to convince, obtain, and accumulate power and that, currently, is considered an equally efficient form of building power as the traditional *hard power*. In the current international context, STI is not only understood as a technical and secondary element, but rather as a substantial instrument to increase the power of people, institutions, and countries.

– *Mechanism of social innovation*: A third role taken on by STI is its use as a generator of ideas and creative processes that allow for the resolution of global problems and challenges within the new global agenda. Essentially, scientific knowledge is considered as a useful tool to face and give a concrete reply to the multiple common political, social, economic, environmental, and cultural issues faced by humanity. This involves the production, governance, and creative application of know-how and scientific knowledge as the solution of structural problems such as poverty, inequality, lack of food, machines, and medicines, the prevention of natural disasters or catastrophes, among many other topics in the current global agenda.

– *Democratizing element:* Currently, knowledge can be considered as a new factor in the diffusion and consolidation of more democratic public spaces, as STI development allows an increase in individuals' empowerment. The massification of digital information and communication media and individual electronic devices are allowing a larger access, participation, and empowerment and a more active role to many citizens in the public space, deepening an awareness of the relevance of democratic values for a life in society.

– *Factor for strategic-military development*: Technological advances have throughout history had an important function in the development of the international actors' defense and security apparatus and have been a great benefit for those that have known how to use technological advances to their own advantage. Currently, this traditional role has intensified due to new scientific and technological discoveries that are especially revolutionizing the military field. Changes observed are multiple and deep and allow us to think that the wars of the future will have a very different format to the one we know currently as a consequence of the application of the latest technological advances both in soldiers and armaments and in virtual fights. Indeed, all actors that know how to most efficiently develop, adapt to, and use new technologies in their military apparatus will obtain significant advantages in the future geopolitical international scenery.

The consolidation of these five roles of STI in the international system allows seeing that, in recent decades, scientific knowledge has become an increasingly important factor in international relations and that, in the current international context, it must be thought of as a decisive element of international dynamics.

Global configuration of international scientific relations

The new roles assumed by STI in the field of international studies have allowed us to understand that the links established between STI and international actors, phenomena, and processes originate and develop within a very precise subfield of action: *international scientific relations*. Aside from being a delimited empirical and factual space, in this project, it has been assumed since the beginning that ISR should be understood also as a subfield of study and/or subdiscipline within international studies where the empirical phenomena linking STI with international relations are specifically studied.

The use of the systemic perspective in the research has allowed us to understand the space comprising ISR as a system unto itself, of which it is possible to know both its parts and its global configuration. For this, to begin with (Part 2: Analytical Framework) the research has allowed to identify, describe, and study some of the most relevant systemic parameters of ISR: the *international context* where it can be found; the *actors* that intervene, the *relations* and *interactions* established, and the internal *processes* generated within them. As a result of this analysis, it is possible to count a few significant discoveries:

– *International context*: ISR and STI exist within an international context that contains and conditions them, and with which they establish constant interactions. The changes that have taken place in the international system in recent decades in the political, economic, and social domains have had a special effect and impact on the role played by scientific knowledge and in the way that ISR is configured. Among these contextual factors, the following stand out: (i) the end of the Cold War period and the modification of the bipolar logic; (ii) changes in the international political system and the transformation of the role of the Nation-State; (iii) emergence of a new period in the capitalist economic system, privileging scientific knowledge as an essential input; (iv) rise in the globalization, which expands the links between actors in the international system; (v) the acceleration and intensification of a scientific and technological revolution with extraordinary systemic impact; and (vi) the realization that we are passing through a historic moment of transition (intersystemic transition) to a new world order.

- *Actors*: It is possible to identify and describe a multiplicity of international actors related to STI within ISR, which stand out because of (i) a considerable increase in the number of State and non-State actors, (ii) a larger and deeper interest from all actors in STI, (iii) the appearance of new State and non-State actors, (iv) the reconfiguration of traditional actors that now seek to adapt to the new reality by changing their strategies to more actively take part in STI processes, and (v) the conformation of a new *pluricentric* or *polycentric* knowledge system created by new and multiple State and non-State actors.
- *Interaction and relations*: The main interaction dynamics and relations within ISR reveal that these links have increased, accelerated, deepened, and intensified, acquiring new characteristics. Among them are (i) a larger quantity, diversity, and modality of interactions that involve more actors; (ii) conflictive links, which have extended to topics such as property rights, patents, or the fight for world *talent*, are maintained and strengthened; (iii) the emergence of new spaces, instruments, and cooperative interaction modalities such as STI diplomacy, the *Triple Helix* model or the expansion of interuniversity cooperation; (iv) the competitive logic is installed between the State and non-State actors through an increase in the competitive *race* between them; and (v) asymmetric interactions are maintained and deepened, which in turn widen the *cognitive divide* and *dependency*.
- *Processes*: Five internal processes and/or operative mechanisms of scientific knowledge have been identified, which explain all of the procedures that STI undergoes from its creation to its final application. These processes (production, intermediation, distribution, application, and governance) are not linear or monocausal, but, on the contrary, overlap and are interactive and multicausal, which turns STI into a very complex process.
- The STI *production* processes have new knowledge creation methods that involve more actors, interactions, places, objectives, procedures, methodologies, and evaluation methods, at the same time that R&D investments and expenditures from all international actors increase.
- *Intermediation* mechanisms have been expanded, renewed, and specialized according to the goals: the transmission through formal education has sped up and intensified; mechanisms for the transfer of knowledge and technology to the productive sector are deepened; the ways of diffusing STI through electrical and digital media are renewed; and the mobility and circulation of highly skilled workers, who personally transport and transmit scientific knowledge throughout the international system, is encouraged.
- The *geopolitical and geoeconomic distribution* of scientific knowledge is still unequal and concentrated both in certain actors and geographical places.

- The *application* processes show important innovations with regards to where and how knowledge is destined (applied research, economic development, the defense sector, engineering, technology and health, and a strong commitment with new emerging technologies), and how this same knowledge is used (economic sector over other social goals).
- Lastly, we must consider the new mechanisms that have emerged for the planning, organization, and execution of STI within the international system and that have transformed the phenomenon into a new topic of *governance*. Currently, the new *governance of STI* is a part of a much more complex process where more actors play a role, through a great variety of interactions carried out in multiple levels and acting arenas simultaneously (local, national, regional, and international).
- *Global configuration*: Complementary to these results, later (Part 3: Explicative Framework), the systemic research has allowed us to reach accurate conclusions and explanations of the global configuration that ISR has adopted in the 21st century. The most revealing findings are the following:
- The global configuration of ISR is openly *unequal* as a result of a *concentrated, hierarchical, segmented,* and *sectorized distribution* of STI both in the geographic location as well as in actors:
    o The *geopolitical* and *geoeconomic concentration* is notorious between countries (the United States, China, the European Union), cities (major capitals), regions (North America, the European Union, and East Asia), and within the same region (strong internal differences in all continents). There is also a centralization in the distribution of world-class universities and international students' destinations (the United States, the United Kingdom, Canada), think thanks and NGOs (the United States and the United Kingdom), and transnational companies (the United States, the European Union, and China).
    o At the same time, a great *concentration of actors* can also be observed, whether States (the United States, China, and the European Union), cities (major capitals and innovative places), universities (world class), companies (those listed in the financial sector), or NGOs and think thanks (through rankings).
    o Last, inequality in the new global structure of ISR is equally perceivable in the increase of the knowledge *hierarchization* and *segmentation* processes, which extend to States (developed, developing, and poor), regions (North America, the European Union, and East Asia), cities (innovation hubs), companies (based on their size, innovation capacity,

and access to markets), universities (rankings), and NGOs and think tanks (according to their funding sources and efficiency in consulting).

- The global configuration of ISR is *multipolar* as a consequence of all the polarization processes that are currently taking place in the international system between actors linked to STI (between State actors, between State and non-State actors, and between non-State actors), which show a clear and constant trend towards a larger numerical quantity of actors interested and with the capacity of acting and influencing in the process:

    o The process of polarization between States shows a transitioning global structure, from an almost exclusive hegemony of the United States and, on a smaller scale, the historical *triad* of scientific knowledge to a more extended, distributed, and global configuration between countries that now includes China, but also other emerging countries such as India, South Korea, Russia, Brazil, and South Africa.
    o At the same time, the processes of polarization between State and non-State actors have also changed due to the apparition of new international actors, such as companies, think tanks, IGOs or NGOs, which show interest and the capacity of influencing in STI and which have been able to balance and challenge the quasi-monopoly States and universities had on knowledge.
    o Lastly, non-State actors also take part in a power rebalancing process between themselves as a consequence of the strong competitive race in which universities, think tanks, NGOs, epistemic communities, and companies have started among them.

- The global configuration of ISR is *dynamic, unstable*, and *changing* due to the phenomenal increase in the quantity, intensity, and variability of actors, relations, processes, topics, and phenomena that are now part of the global structure of STI. This extraordinary acceleration of the internal dynamics of ISR has led to a multiplicity of new emerging realities that emerge from within it as a consequence of the constant interactions between international actors, which become strong macrotrends that affect ISR itself and the international system as a whole. Among them are the following:

    o The formation of *knowledge divides* (digital, cognitive, and scientific).
    o The increase in the *marketing* and *commercialization* of STI.
    o A new *global* and *multilevel governance* of STI.

o The persistence of the *gender divide*, despite the improvement of women's status in STI.
o A slow but sustained *geopolitical* and *geoeconomic shift* of STI, from the West to East.
o The emergence of new *areas in science and technology*, with incalculable relevance and impact for the international system.
o The emergence of a *virtual world* opens renewed possibilities for cooperation, but also for conflict.

International scientific relations and international system

After discovering the role of scientific knowledge in the 21st century's international system and analyzing the global configuration of ISR, it is possible to state three crucial findings with regards to the impact that ISR have on the international system as a whole:

(i) ISR receives strong demands from the international system in general, and in particular from other subsystems, which are processed internally and to which they later give a reply in the form of feedback towards each of the subsystems and the international system as a whole.

(ii) ISR has a special impact in some of the subsystems composing the international context; in particular, we highlight the influence that STI has in the economic subsystem (as a strategic resource for innovating, competing, and obtaining economic benefits), in the political subsystem (as an instrument of power and democratizing element), in the strategic-military subsystem (as a tool for military development) and in the social subsystem (as a mechanism of social innovation).

(iii) ISR feed back into and self-regulate the international system through trends generated within them, which end up contributing to the forces of change and/or continuity that impact the stability and balance of the international system as a whole through *continuity macrotrends* (empowerment of traditional actors, bolstering of the State's active role, preservation of inequality gaps, the persistence of asymmetric relations, etc.) and through *change macrotrends* (emergence of new actors, new modes of interaction, new forms of governance, new roles of knowledge, etc.).

Based on this analysis, it is possible to reach two final conclusions with regards to the place STI and ISR hold within the 21st century's international system:

➢ On the one hand, it is possible to state that *scientific knowledge* has become a core element of the international system because of two main reasons:

- o   *It is the most important strategic resource in the current world order.*
- o   *It has become the new ordering principle of current international relations.*

➢ On the other hand, it is possible to state that *international scientific relations* are impacting and modifying the 21st century's international system in two essential ways:
- o   *Substantially changing the main systemic parameters, that is, the essential parts that form the international system (actors, relations, agenda topics, processes, dynamics, phenomena, etc.).*
- o   *Modifying the global structure and configuration of the system (new polarizations, divides, hierarchies, resources of power, etc.).*

Epilogue

Through these findings, we have discovered the existence of a renewed, complex, and powerful global agenda linked to ISR which, due to its current relevance and future impact, opens a wide range of implications for all actors in the international system. In concrete terms, research shows the necessity of working on public policies, private strategies, and global governance mechanisms that include all actors in all levels of actuation, that stimulate cooperation mechanisms, that reduce conflicts, and that take into consideration a more equitable distribution of the new knowledge created. Science, technology, and innovation can become the most transcendental and revolutionary elements of the international system if we manage to make technical advances and progress benefit humanity as a whole, but it can also become a dangerous tool that can further existing inequalities. The present study has allowed us to thoroughly understand that the answer to this issue will greatly depend on the capacity of the main actors of the international system to tackle the agenda in a joint, cooperative, and responsible way.

At the same time, in the academic context, research shows the necessity of new approaches which, from multiple areas of study and in disciplinary and interdisciplinary ways, analyze the new agenda linked to ISR. The major changes that have taken place in the international system in the years following the Cold War have forced us to reopen an academic debate that would give concrete explanations of the new characteristics that the new world order is acquiring. The purpose of this research is to promote the study and the analysis of new scientific projects specifically linked with the subdiscipline of ISR and the role of STI in the international system, as it is understood as being one of the most relevant subjects of study in the current global international agenda.

In this process, it is necessary to renew the theoretical-methodological scaffolding of international studies to be able to offer new responses bearing in mind new empirical phenomena. For this, the usage of approaches and paradigms such as the systemic perspective can give promising theoretical and methodological tools to analyze the international reality in a different way. The usage of systemic analysis models as instruments for the study and understanding of international phenomena can also be considered to be a useful method to recreate and represent reality and to offer, through this modeling, rigorous scientific explanations. It is also necessary to promote the formulation of more scientific projects that opt for a methodological strategy incorporating a multi- and interdisciplinary approach, due to the increasing interrelation and complexity of the current international system, which forces us to use and combine methodological tools stemming from multiple disciplines to give more academic responses that are more complete and faithful to the reality studied.

Lastly, it is understood that the findings presented in this book are only the starting point for new research projects that will expand on and deepen what has been discovered here. *International scientific relations* are awaiting a wider and deeper approach in many of the new realities that have been revealed in this research, whether phenomena (STI diplomacy, diasporas, emerging technologies, etc.), geographical areas (East Asia, the European Union, emerging countries, etc.), actors (sub-State entities, epistemic communities, or think tanks), processes (production, intermediation or governance), interactions (cooperation, conflict, competition, etc.), or emerging realities (divides, virtual world, geopolitical changes, etc.). All these from theoretical and methodological frameworks that would give new revelations and solutions to the current international agenda of ISR.

All these considerations and reflections are the direct result of extensive scientific research carried out by the author for more than fifteen years on the particular intersection between science, technology, and innovation, and international relations which, finally, have been reflected in this book. The main conclusions speak about new phenomena, actors, dynamics, processes, emerging realities, configurations, and roles within the framework of contemporary ISR. In the 21st century, the social and scientific need for work that addresses the link between STI and international relations is more relevant than ever, so it is hoped that the findings of this work will prove to be a useful contribution and, at the same time, pave the way for new research that will stimulate empirical study, theoretical analysis, academic debate and practical applications of an international phenomenon that is as complex as it is unexplored.

REFERENCES

Adkings, S. S. (2016). "The 2016–2021 Worldwide Self-paced eLearning Market: Global eLearning Market in Steep Decline." *Ambient Insight.* August.
Adler, E., and P. M. Haas. (2009). "Conclusion: Epistemic Communities, World Order, and the Creation of a Reflective Research Program." *International Organization*, vol. 46, no. 1, Knowledge, Power, and International Policy Coordination (inter 1992), pp. 367–90..
Albornoz, M. (1997). "La política científica y tecnológica en América Latina frente al desafio del pensamiento único." *Redes*, vol. 4, no. 10, octubre 1997, pp. 95–115.
———. (2001). "Política Científica y Tecnológica. Una visión desde América Latina." *Revista CTS*, no. 1, septiembre–diciembre 2001.
———. (2007). "Los problemas de la ciencia y el poder." *Revista CTS*, vol. 3, no. 8, pp. 43–65.
Altbach, P. (2006). "Chapter 8: Globalization and the University: Realities in an Unequal World." In *International Handbook of Higher Education*, ed. J. Forest and P. Altbach. Dordrecht: Springer.
Altbach, P., L. Reisberg, and L. Rumbley. (2010). "Tracking a Global Academic Revolution." *Change*, vol. 42, no. 2, March/April, pp. 30–39.
Ander-Egg, E. (2001). *Métodos y técnicas de investigación social.* Buenos Aires: Lumen.
———. (2010). *Metodologías de acción social.* Buenos Aires: Lumen-Humanitas.
Arocena, R. (2007). "Sobre la democratización de la ciencia y la tecnología." *Quantum*, vol. II, no. 1, octubre 2007, pp. 7–14.
———. (2018). "A Prospective Approach to Learning- and Innovation-based Development." *Millennial Asia*, vol. 10, no. 2, pp. 127–47.
Baker, D. (2020). *Technology, Patents, and Inequality: An Explanation that Even Economists Can Understand.* Center for Economic and Policy Research. January 29.
Bauman, Z. (1998). "On Glocalization: or Globalization for some, Localization for some Others." *Thesis Eleven* vol. 54, no. 1, August 1, pp. 37–49.
———. (2000). *Liquid Modernity.* London: Polity Press.
Bell, D. (1973). *The Coming of Post-industrial Society: A Venture in Social Forecasting.* New York: Basic Books.
Bourdieu, P. (2004). *Science of Science and Reflexivity.* Cambridge: Polity Press.
Breton, G., and M. Lambert. (2003). *Universities and Globalization: Private Linkage, Public Trust.* Paris: UNESCO.
Brunner, J. J. (2010). "Globalización de la Educación Superior: Critica de su figura ideológica." *RIES, Revista Iberoamericana de Educación Superior*, vol. I, no. 2, pp. 75–83.
Bull, H. (1966). "International Theory: The Case for a Classical Approach." *World Politics*, vol. 18, no. 3, pp. 361–77.
Bunge, M. (1972). *La Ciencia, su Método y su Filosofía.* Buenos Aires: Siglo Veinte.

———. (1979). *Treatise on Basic Philosophy. Volume 4: Ontology II: A World of Systems*. Dordrecht: D. Reidel.

———. (1980a). *Epistemología*. Barcelona: Ariel.

———. (1980b). *La investigación científica. Su estrategia y su filosofía*. Barcelona: Ariel.

———. (1996). *Finding Philosophy in Social Science*. NH: Yale University Press.

Bunge, M. (2003). *Emergence and Convergence: Qualitative Novelty and Unity of Knowledge*. Toronto: University of Toronto Press.

———. (2009). "Dos enfoques de la Ciencia: Sectorial y Sistémico." *Rev. Real Academia de Ciencias. Zaragoza*, vol. 64, pp. 51–63.

Bush, V. (1945). *Science: the Endless Frontier*. Washington, DC: United States Government.

Caballero, S. (2009). *Comunidades epistémicas en el proceso de integración sudamericana*. (Seminario de investigadores en formación (SIF-UAM), Departamento de Ciencia Política y Relaciones Internacionales, UAM).

Calame, P. (2014). "Rethinking Decentralisation and Strengthening the Participation and Decision-Making Powers of Citizens in the Territories." *Connexions*, vol. 101, no. 1, pp. 115–24.

Calduch Cervera, R. (2017). *Métodos y técnicas de investigación en Relaciones Internacionales*. (Curso de Doctorado, Universidad Complutense de Madrid). Madrid: Universidad Complutense de Madrid.

Campanario, S. (2017). "Geopolítica de la innovación, un juego de países y empresas." *La Nación – Economía*. 22 de octubre de 2017.

Carayannis, E. G., and D. F. J. Campbell. (2006). "'Mode 3': Meaning and Implications from a Knowledge Systems Perspective." In *Knowledge Creation, Diffusion, and Use in Innovation Networks and Knowledge Clusters: A Comparative Systems Approach across the United States, Europe, and Asia*, 1–25. Westport, CT: Praeger Publishers.

———. (2009). "'Mode 3' and 'Quadruple Helix': toward a 21st-Century Fractal Innovation Ecosystem." *International Journal of Technology Management* vol. 46, no. 3/4, pp. 201–34.

Carayannis, E. G., D. F. J. Campbell, and S. S. Rehman. (2016). "Mode 3 Knowledge Production: Systems and Systems Theory, Clusters and Networks." *Journal Innovation and Entrepreneurship*, vol. 5, 17.

Castells, M. (1996). *The Rise of the Network Society*. Oxford: Blackwell Publishers.

Castells, M. (2000). *Internet y la Sociedad Red*. (Lliçó inaugural del programa de doctorat sobre la societat de la informació i el coneixement. Universitat Oberta de Catalunya (UOC), Barcelona, Espanya, octubre 2000).

Castells, M. (2005). *Network Society: From Knowledge to Policy*. Washington, DC: Johns Hopkins Center for Transatlantic Relations.

Castells, M. (2009). *Communication Power*. Oxford: Oxford University Press.

Castro, J. (2018). *China ya supera a Estados Unidos en la carrera de la innovación*. Diario Clarin. July 1.

Cervera, J. (2016). *Miras, sensores, drones y robots: la infantería del futuro ya está aquí*. El Confidencial. 7 de marzo de 2016.

Chaminade, C., and B.-Å. Lundvall. (2019). "Science, Technology, and Innovation Policy: Old Patterns and New Challenges." In *Oxford Research Encyclopedia, Business, and Management*. May.

Comisión Europea. (2010). *EUROPA 2020. Una estrategia para un crecimiento inteligente, sostenible e integrador*. Comunicación de la Comisión Europea, Bruselas, March 3, . Bruselas: Comisión Europea.

Congressional Research Service. (2020). *Global Research and Development Expenditures: Fact Sheet*, updated on April 29.
Cox, R. (1993). "Chapter 2: Gramsci, Hegemony, and International Relations: An Essay in Method." In *Gramsci, Historical Materialism, and International Relations,* ed. S. Gill, pp. 49–65). New York: Cambridge University Press.
———. (1996). *Approaches to World Order.* New York: Cambridge University Press.
Dagnino, R., and H. Thomas. (1999). "Latin American Science and Technology Policy: New Scenarios and the Research Community." *Science, Technology and Society*, vol. 4, no. 1, pp. 35–54.
Dallanegra Pedraza, L. (1998). *El Orden Mundial del Siglo XXI.* Buenos Aires: Ediciones de la Universidad.
Dallanegra Pedraza, L. (2010). "Teoría y metodología de la geopolítica: Hacia una geopolítica de la construcción de poder." *Revista mexicana de ciencias políticas y sociales*, no. 210, pp. 15–44.
Dallanegra Pedraza, L. (2012). "Escenarios sobre el orden internacional." *Reflexión política*, vol. 14, no. 28.
Dedijer, S. (1968). Early Immigration. In *The Brain Drain*, (coord.) W. Adams. New York: MacMillan Co.
del Arenal Moyúa, C. (2009). *Mundialización, creciente interdependencia y globalización en las relaciones internacionales.* (Cursos de Derecho Internacional y Relaciones Internacionales de Vitoria-Gasteiz – 2008). 182–268. Leioa: UPV/EHU.
Del Canto Viterale, F. (2019). "Developing a Systems Architecture Model to Study the Science, Technology, and Innovation in International Studies." *Systems*, vol. 46, no. 7.
de Semir, V. (2003). "Medios de comunicación y cultura científica." *Quark: Ciencia, medicina, comunicación y cultura*, nos. 28–29.
de Sousa Santos, B. (2005). *La Universidad en el Siglo XXI. Para una reforma democrática y emancipadora de la universidad.* Buenos Aires: Laboratorio de Políticas Públicas y Miño y Davila.
———. (2015). *Epistemologies of the South: Justice against Epistemicide.* New York: Routledge.
———. (2018). *The End of the Cognitive Empire: The Coming of Age of Epistemologies of the South.* Durham: Duke University Press.
de Sousa Santos, B., and M. P. Meneses. (2019). *Knowledge Born in the Struggle: Constructing the Epistemologies of the Global South.* New York: Routledge.
de Wit, H., and P. G. Altbach. (2020). "Internationalization in Higher Education: Global Trends and Recommendations for Its Future." *Journal Policy Reviews in Higher Education*, vol. 5, pp. 28–46.
Der Derian, J., and M. J. Shapiro. (1989). *International/Intertextual Relations: Postmodern Readings of World Politics.* Lexington, MA: Lexington Books.
Derrida, J. (1974). *Of Grammatology.* Baltimore: Johns Hopkins University Press.
Deudney, D. (2018). "Turbo Change: Accelerating Technological Disruption, Planetary Geopolitics, and Architectonic Metaphors." *International Studies Review*, vol. 20, pp. 23–31.
Drucker, P. (1969). *The Age of Discontinuity: Guidelines to Our Changing Society.* New York: Harper and Row.
———. (1992). *The Age of Discontinuity: Guidelines to Our Changing Society.* New York: Harper and Row.
Dumont, J. C., and G. Lemaitre. (2005). *Counting Immigrants and Expatriates in OECD Countries: A New Perspective.* (OECD Social, Employment, and Migration Working Papers

no. 25. UN/POP/PD/2005/09). New York: United Nations Expert Group Meeting on International Migration and Development.

Echeverría, J. (2008). "Transferencia de conocimiento entre comunidades científicas." *ARBOR Ciencia, Pensamiento y Cultura*, vol. CLXXXIV, no. 731, mayo-junio, pp. 539–48.

———. (2014). *Innovation and Values: A European Perspective*. Reno: UNR–CBS.

Etzkowitz, H., and Leydesdorff, L. (2000). "The Dynamics of Innovation: From National Systems and 'Mode 2' to a Triple Helix of university-industry–government relations." *Research Policy*, vol. 29, no. 2, pp. 109–123

Euroasia Group. (2018). *Top Risk 2018*. Report, New York. January 2.

European Commission. (2001). *European Governance: A White Paper*. COM 428, July 25. Brussels: European Commission.

———. (2004). *Europe needs more Scientifics. Increasing Human Resources for Science and Technology in Europe*. Report of the High-Level Group on Human Resources for Science and Technology in Europe 2004. Brussels: European Commission.

———. (2013). *Responsible Research and Innovation (RRI), Science, and Technology*. Special Eurobarometer 401. November.

European Council. (2010). *Project Europe 2030. Challenges and Opportunities*. A report to the European Council by the Reflection Group on the Future of the EU 2030. May.

European Strategy Forum on Research Infrastructures – (ESFRI). (2019). *Strategy Report on Research Infrastructures. Roadmap 2021 Public Guide*. September 25. Brussels: European Commission.

European Union. (1999). *The Bologna Declaration of 19 June 1999*. Joint declaration of the European Ministers of Education.

———. (2009). *The Lund Declaration*. July. Lund: European Union.

Faist, T. (2005). "Espacio social transnacional y desarrollo: una exploración de la relación entre comunidad, estado y mercado." *Migración y Desarrollo*, segundo semestre, pp. 2–34.

Fedoroff, N. (2009). "Science Diplomacy in the 21st Century." *Cell*, vol. 136, no. 11, January 9, pp. 9–11.

———. (2011). "21st-Century Challenges Require Global Focus by Scientists." *Science*, vol. 331, no. 6016, pp. 422–25.

Ferrero, M., and I. Filibi López. (2006). "¡Bárbaros en Delfos! Geopolítica del Conocimiento y Relaciones Internacionales ante el siglo XXI." *CONfines*, 2/3, enero-mayo 2006, pp. 27–44.

Flink, T., and U. Schreiterer. (2010). "Science Diplomacy at the Intersection of S&T Policies and Foreign Affairs: Toward a Typology of National Approaches." *Science and Public Policy*, vol. 37, no. 9, November, pp. 665–77.

Flink T., and N. Rüffin. (2019). "The Current State of the Art of Science Diplomacy." In *Handbook on Science and Public Policy*, ed. S. Simon, S. Kuhlmann, W. Canzler, and J. Stamm, pp. 104–21. Cheltenham: Edward Elgar.

Fojon, E. (2018). "La cuarta revolución industrial, el "algoritmo de la guerra" y su posible aplicación a la defensa española." *Real Instituto Elcano*. 9 de marzo de 2018.

Foray, D. (2006). *The Economics of Knowledge*. Cambridge: MIT Press.

Forest, J., and P. Altbach. (2006). *International Handbook of Higher Education*. Dordrecht: Springer.

Foucault, M. (1972). *The Archeology of Knowledge*. New York: Pantheon Books.

Freeman, C. (1995). "The 'National System of Innovation' in Historical Perspective." *Cambridge Journal of Economics*, vol. 19, no. 1, February, pp. 5–24.

Friedrichs, J. (2001). "The Meaning of New Medievalism." *European Journal of International Relations*, vol. 7, no. 4, December, pp. 475–501.

Fukuda, K. (2019). "Science, Technology and Innovation Ecosystem Transformation toward Society 5.0." *International Journal of Production Economics*, vol. 220, February.

Fundación Innovación Bankinder. (2011). "La educación del siglo XXI." *Una apuesta a futuro*. Número 16, noviembre 2011.

———. (2016). *Ciberseguridad, un desafío mundial*. Número 25, mayo 2016.

Funtowicz, S., and J. Ravetz. (2003). *Post-Normal Science*. International Society for Ecological Economics and Internet Encyclopaedia of Ecological Economics. February.

García Guadilla, C. (1996). *Conocimiento, educación superior y sociedad en América Latina*. Caracas: Nueva Sociedad.

———. (2005). Complejidades de la globalización e internacionalización de la educación superior: interrogantes para América Latina. *Cuadernos del Cendes*, ISSN 1012-2508, no. 58 (Enero-Abril),pp. 1–22

———. (2010). *Educación Superior Comparada. El Protagonismo de la Internacionalización*. Caracas: UNESCO - CENDES.

Gibbons, M., C. Limoges, E. Nowotny, S. Schwartzman, P. Scott, and M. Trow. (1994). *The New Production of Knowledge: The Dynamics of Science and Research in Contemporary Societies*. London: SAGE Publications.

Gilman, D. (2010). *The New Geography of Global Innovation*. Report Global Market Institute, Goldman Sachs Global Investment Research, September 20. Goldman Sachs Group.

Global Geneva. (2018). Top 500 NGOs.

Gobierno Vasco. (2019). *PCTI EUSKADI 2030. Líneas Estratégicas y Económicas Básicas*. Vitoria-Gasteiz: Gobierno Vasco.

Godin, B. (2003). *The Most Cherished Indicator: Gross Domestic Expenditure on R&D (GERD)*. Project on History and Sociology of S&T Statistics, Working Paper no. 22. Montreal: Canadian Science and Innovation Indicators Consortium.

Goldin, I. (2016). *Cómo la inmigración ha cambiado y mejorado el mundo*. World Economic Forum. 21 enero 2016.

González de la Fe, T. (2009). "El modelo de Triple Hélice de relaciones Universidad, Industria y Gobierno: un análisis crítico." *ARBOR Ciencia, Pensamiento y Cultura*, CLXXXV, 738, julio-agosto 2009, pp. 739–55.

Haas, P. (1992). "Introduction: Epistemic Communities and International Policy Coordination." *International Organization*, vol. 46, no. 1, pp. 1–35.

Haass, R. (2017). *A World in Disarray: American Foreign Policy and the Crisis of the Old Order*. New York: Penguin Books.

———. (2020). *Deglobalization and Its Discontents*. Project Syndicate. May 12.

Habermas, J. (1968). *Knowledge and Human Interest*. Cambridge: Polity Press.

———. (2000). "Crossing Globalization's Valley of Tears." *New Perpective Quarterly*. Fall 2000.

Halliday, F. (2000). "Global Governance: Prospect and Problems." *Citizenship Studies*, vol. 4, no. 1, pp. 19–33.

Halliday, F., N. Zúñiga García-Falces, B. Wang. (2006). *Las Relaciones Internacionales y sus debates*. Madrid: FUHEM, Centro de Investigación para la Paz.

Held, D. (2000). "Regulating Globalization? The Reinvention of Politics." *International Sociology*, vol. 15, no. 2, June, pp. 394–408.

Hirst, P., and G. Thompson. (2002). "The Future of Globalization." *Cooperation and Conflict: Journal of the Nordic International Studies Association*, vol. 37, no. 3, pp. 247–65.

Hooghe, L., and G. Marks. (2001). Multi-Level Governance and European Integration. Lanham: Rowman & Littlefield.

———. (2020). "A Postfunctionalist Theory of Multilevel Governance." *The British Journal of Politics and International Relations*, vol. 22, no. 4, pp. 820–26.

Innerarity, D. (2011). *La democracia del conocimiento. Por una sociedad inteligente*. Barcelona: Paidós.—

———. (2013). "Power and Knowledge: The Politics of the Knowledge Society." *European Journal of Social Theory*, vol., 16, no. 1, pp. 3–16.

Innovation Cities Program. (2019). "Innovation Cities Index 2019—Global City Rankings by 2thinknow." November 11, www.innovation-cities.com/index-2019-global-city-rankings/18842/.

Internet World Stats. (2020). "World Internet Users Statistics and 2021 World Population Stats." www.internetworldstats.com/stats.htm.

Jalan, K., and E. L. Kopchia. (2020). "15 Best Science and Technology Research Labs in the World." *RankRed*, 11 April. www.rankred.com/best-science-and-technology-research-labs/.

Jarvis, P. (2001). *Universities and Corporative Universities. The Higher Education Learning Industry in Global Society*. London: Kogan Page.

Kaeser, J. (2018). *The World Is Changing. Here's How Companies Must Adapt*. World Economic Forum. January 25.

Kaplan, M. (1957). *System and Process in International Politics*. New York: John Wiley and Sons.

Keidanren (Japan Business Federation). (2016). *Toward the realization of the new economy and society. Keidanren: Outline. Reform of the Economy and Society by the Deepening of Society 5.0*. April 19.

Kennedy, P. (1987). *The Rise and Fall of the Great Powers: Economic Change and Military Conflict from 1500 to 2000*. New York: Vintage Books.

Keohane, R., and J. Nye. (1987). "Power and Interdependence Revisited." *International Organization*, vol. 41, no. 4, pp. 725–53.

———. (1998). "Power and Interdependence in the Information Age." *Foreign Affairs*, vol. 77, p. 81.

King, R. (2004). *The University in the Global Age*. Basingstoke: Palgrave MacMillan.

Knight, J. (2006). "Internationalization: Concepts, Complexities, and Challenges." In *International Handbook of Higher Education*, ed. J. Forest and P. Altbach. chap. 11, vol. 18. Dordrecht, the Netherlands: Springer.

———. (2018). "International Education Hubs." In *Geographies of the University*, ed. P. Meusburger, M. Heffernan, L. Suarsana. chap. 21, pp. 637–55. Springer Open.

Kolo, P., R. Strack, P. Cavat, R. Torres, and V. Bhalla. (2013). *Corporate Universities: An Engine for Human Capital*. Boston Consulting Group. July.

Krepinevich, A. (2011). "Get Ready for the Democratization of Destruction." *Foreign Policy*. August 15.

Kuhlmann S., and A. Rip. (2018). "Next-Generation Innovation Policy and Grand Challenges." *Science and Public Policy*, vol. 45, no. 4, pp. 448–54.

Kuhn, T. S. (1962). *The Structure of Scientific Revolutions*. Chicago: University of Chicago Press.

Kukso, F. (2010). *El renacer de la Big Science*. Le Monde diplomatique, no. 132.

Lamo de Espinosa, E., J. M. González García, and C. Torres Albero. (1994). *Sociología del conocimiento y de la ciencia*. Madrid: Alianza.

Lander, E. (2008). "La ciencia neoliberal." *Tabula Rasa*, no. 9, julio–diciembre 2008, pp. 247–83.

Leijten, J. (2019). "Science, Technology and Innovation Diplomacy: A Way Forward for Europe." *Institute for European Studies Policy*, brief issue 2019/15.

Leydesdorff, L. (2012). "The Triple Helix of University-Industry-Government Relations." In *Encyclopedia of Creativity, Innovation, and Entrepreneurship*, ed. E. Carayannis and D. Campbell. New York: Springer.

Li, C., and F. Lalani. (2020). *The COVID-19 Pandemic Has Changed Education Forever: This Is How*. World Economic Forum. April.

Lijesevic, J. (2010). *Science Diplomacy at the Heart of International Relations*. E-International Relations. April 1.

Luchilo, L. (2006). "Movilidad de estudiantes universitarios e internacionalización de la educación superior." *Revista CTS*, vol. 3, núm. 7, septiembre de 2006, pp. 105–33.

Lumina Foundation. (2018). *Tracking America's Progress Toward 2025*. A Stronger Nation 2017. https://www.luminafoundation.org/stronger-nation/report/2020/#nation.

Lundvall, B. A. (1985) "Product Innovation and User-Producer Interaction, Industrial Development." *Research Series 31*. Aalborg: Aalborg University Press.

———. (2004). *National Innovation Systems–Analytical Concept and Development Tool*. Paper presented at the DRUID Tenth Anniversary Summer Conference 2005 on Dynamics of Industry and Innovation: Organizations, Networks, and Systems, Copenhagen, Denmark, June 27–29, 2005.

Luttwak, E. (1990). "From Geopolitics to Geo-Economics: Logic of Conflict, Grammar of Commerce." *The National Interest*, no. 20 (Summer 1990), pp. 17–23.

Makri, A. (2013). *Science and NGO Practice: Facts and Figures*. SciDevNet. May 22.

Marcuse, H. (1964). *One-Dimensional Man: Studies in the Ideology of Advanced Industrial Society*. Boston: Beacon Press.

Mariscal, N. (2006). "Gobernanza múltiple y plural." *Europa Euskadi. Red Vasca de Información Europea (REVIE)*, vol. 2006, no. 191, pp. 5–6.

———. (2011). "La contribución de la Unión Europea a la gobernanza global." *Cuadernos europeos de Deusto*, no. 45 (Ejemplar dedicado a: La contribución de la Unión Europea a la gobernanza global), pp. 13–19.

———. (2017). "La Unión Europea en la Gobernanza Global." *Cuadernos europeos de Deusto*, no. 56 (Ejemplar dedicado a: Governing Mobility in Europe: Interdisciplinary Perspectives), pp. 181–204.

Martinage, R. (2014). *Toward a new offset strategy. Exploiting US long-term advantages to restore US global power projection capability*. Washington DC: Center for Strategic and Budgetary Assessments (CSBA).

Mayer, M., M. Carpes, and R. Knoblich. (2014). *The Global Politics of Science and Technology Vol. 1*. Berlin: Springer.

Mayer, M., and M. Acuto. (2015). "The Global Governance of Large Technical Systems." *Millennium*, vol. 43, pp. 660–83.

Mc Afee and CSIS. (2018). *Economic Impact of Cybercrime—No Slowing Down*. February. https://www.csis.org/analysis/economic-impact-cybercrime.

Mc Gann, J. (2019). *The Global Go to Think Tanks Index Report*. Philadelphia: University of Pennsylvania.

Meek, V., U. Teichler, and M. L. Kearney, eds. (2009). *Higher Education, Research, and Innovation: Changing Dynamics*. Report on the UNESCO Forum on Higher Education, Research and Knowledge 2001–2009. International Centre for Higher Education Research Kassel, University of Kassel, Germany.

Meyer, J. B., D. Kaplan, and J. Charum. (2001). "Scientific Nomadism and the New Geopolitics of Knowledge." *International Social Science Journal*, vol. 53, no. 168, pp. 309–21.

Mignolo, W. (2003). "Globalization and the Geopolitics of Knowledge: The Role of the Humanities in the Corporate University." *Nepantla: Views from South 4.1*, vol. 4, no. 1, pp. 97–119.

Mollis, M. (2006). "Capítulo: Geopolítica del saber: biografías recientes de las universidades latinoamericanas." In *Universidad e investigación científica*, ed. H. Vessuri, pp. 85–101. Buenos Aires: CLACSO.

Montuschi, L. (2001). *La economía basada en el conocimiento: Importancia del conocimiento tácito y del conocimiento codificado.* (UCEMA Working Papers: 204). Buenos Aires: Universidad del CEMA.

Morata, F. (2004). *Gobernanza multinivel en la Unión Europea.* Valencia: Tirant lo Blanch.

Moreno Alegre, J. M., and A. Albáizar Fernández. (2008). "La tercera misión de la Universidad." In *Libro Blanco de la Universidad Digital 2010*, ed. J. Laviña Orueta and L. Mengual Pavón (2008), pp. 83-101. Fundación Telefónica, Ariel.

National Science Board and National Science Foundation. (2020). *Science and Engineering Indicators 2020: The State of U.S. Science and Engineering.* NSB-2020-1. Alexandria, VA.

National Science Foundation. (2016). *Info Brief.* NSF 16-317, September.

———. (2018). *Science & Engineering Indicators 2018.* Washington, DC: NSF.

Nonprofit Tech for Good. (2018). *2018 Global NGO Technology Report.* https://assets-global.website-files.com/5da60733afec9db1fb998273/5de6d45aee027c401be467e4_2018-Tech-Report-English.pdf.

Nowotny, H., P. Scott, and M. Gibbons. (2003). "Mode 2 Revisited: The New Production of Knowledge." *Minerva*, vol. 41, pp. 179–94.

———. (2006). "Re-thinking Science: Mode 2 in Societal Context." In *Knowledge Creation, Diffusion, and Use in Innovation Networks and Knowledge Clusters. A Comparative Systems Approach across the United States, Europe and Asia*, ed. E. G. Carayannis and D. F. J. Campbell, pp. 39–51. Westport, CT: Praeger.

Núñez Jover, J., and F. Castro Sánchez. (2005). "Universidad, Innovación y Sociedad: Experiencias de la Universidad de La Habana." *Revista de Ciências da Administração*, vol. 7, no. 13, January/July, pp. 1–21.

Nuñez Jover, J. (2006). "La democratización de la ciencia y el problema del poder." In *La Política: Miradas Cruzadas*, ed. E. Duharte Díaz, E. (Comp.). La Habana: Editorial de Ciencias Sociales.

Núñez Jover, J., H. Ortiz Pérez, T. Proenza Díaz, and A. Rivas Diéguez. (2020). "Políticas de educación superior, ciencia, tecnología e innovación y desarrollo territorial nuevas experiencias, nuevos enfoques." *CTS: Revista iberoamericana de ciencia, tecnología y sociedad*, ISSN 1668-0030, vol. 15, no. 43,pp. 187–208.

Nye, J. (2010). "Prefacio y Capítulo 5: El poder blando y la política exterior americana." *Soft Power, Public Affairs*, New Hampshire, 2004, pp. IX-XIII y 127–47. En: *Relaciones Internacionales*, núm. 14, junio de 2010, pp. 117–140.

———. (2017). "Soft Power: The Origins and Political Progress of a Concept. *Palgrave Communications*, vol. 3, no. 17008.

Observatoire Des Sciences Et Des Techniques (OST). (2016). *Indicateurs de Sciences et de Technologies. Édition 2014.* Rapport de L'Observatoire des Sciences et des Techniques: L'espace mondial, 2014. Paris: OST.

OECD. *Online Database OECD's Library.* https://www.oecd-ilibrary.org/statistics.

———. (1995). *The Measurement of Scientific and Technological Activities: Manual on the Measurement of Human Resources Devoted to S&T: Canberra Manual.* Paris: OECD.

———. (2000). *Knowledge Management in the Learning Society.* Paris: OECD.

———. (2001). *Understanding the Digital Divide*. OECD Digital Economy Papers, no. 49.
———. (2003). *Manual de Frascati*. Paris: FECYT.
———. (2007). *Science, Technology, and Innovation Indicators in a Changing World: Responding to Policy Needs*. Paris: OECD.
———. (2008). "Chapter 4: Assessing the Socio-economic Impacts of Public R&D: Recent Practices and Perspectives." In *OECD Science, Technology, and Industry*, pp. 189–217. Paris: OECD.
———. (2011). *Science, Technology, and Industry Scoreboard 2011: Innovation and Growth in Knowledge Economies*. Paris: OECD.
———. (2012). ANSKILL Database, May. https://stats.oecd.org/OECDStat_Metadata/ShowMetadata.ashx?Dataset=BENCHMARK_STIO&Lang=en&Coords=[INDICATOR].[HC4]&backtodotstat=false.
———. (2013). *Main Science and Technology Indicators*. vol. 2013/1, June. Paris: OECD.
———. (2016). *Science, Technology, and Innovation Outlook 2016*. Paris: OECD.
———. (2017). *OECD Science, Technology, and Industry Scoreboard 2017: The Digital Transformation*. Paris: OECD.
———. (2018). *Oslo Manual 2018. Guidelines for Collecting, Reporting, and Using Data on Innovation*, 4th ed. Paris: OECD.
OECD and EUROSTAT. (2006). *Manual de Oslo. Guía para la recogida e interpretación de datos sobre innovación*. (Tercera Edición, Edición Española). España: Grupo Tragsa.
O'Halloran, D. (2015). *How Technology Will Change the Way We Work*. World Economic Forum. August 13.
Olivé, L. (2005). "Los desafíos de la sociedad del conocimiento: la ciencia, la tecnología y la gobernanza." *Revista Este País*, no. 172, julio 2005, pp. 66–70.
Ortega, A. (2018). "Guerra Fría por la Inteligencia Artificial." *Real Instituto Elcano*. 20 de febrero de 2018.
Ortega, A and F. A. Perez (2018). *A G20 Agenda for Technological Justice*. Real Instituto Elcano. ARI 31/2018–6/3/2018.
Parsons, C. (2015). *What's the Best Policy to Attract High-Skilled Migrants?* World Economic Forum. June 9.
Pellegrino, A. (2001). "Trends in Latin American Skilled Migration: 'Brain Drain' or 'Brain Exchange'?" *International Migration*, vol. 39, no. 5, pp. 111–32.
———. (2004). "Migration from Latin America to Europe: Trends and Policy Challenges." *International Organization for Migration*, no. 16.
Pellegrino, A., J. Bengochea, and M. Koolhaas, eds. (2013). "La migración calificada desde América Latina: tendencias y consecuencias." *Programa de Población, Facultad de Ciencias Sociales*, Universidad de la República.
Perotti, D. E., and R. J. Sanchez. (2011). *La brecha de infraestructura en America Latina y el Caribe*. CEPAL, División de Recursos Naturales e Infraestructura, Santiago de Chile, julio.
Petrash, V., comp. (1998). *Cambio, Contradicción y Complejidad en la Política Internacional del Fin de Siglo*. Caracas: Nueva Sociedad.
Piattoni, S. (2010). *The Theory of Multi-Level Governance: Conceptual, Empirical, and Normative Challenges*. Oxford: University Press.
———. (2018). "Multilevel Governance." In *Handbook of Territorial Politics*, ed. K. Detterbeck and E. Hepburn. Cheltenham: Edward Elgar.
Pike, A., A. Rodriguez-Pose, and J. Tomaney. (2007). "What Kind of Local and Regional Development and for Whom?" *Regional Studies*, vol. 41, no. 9, December, pp. 1253–69.
Polanyi, M. (1969). *Personal Knowledge: Towards a Post-Critical Philosophy*. London: Routledge & Kegan Paul.

Porter, M. E. (1998). "Clusters and the New Economics of Competition." *Harvard Business Review*, November–December, pp. 77–90.

———. (2011). *The Competitive Advantage of Nations: Creating and Sustaining Superior Performance.* New York: Free Press.

PwC. (2018). *The 2018 Global Innovation 1000 Study*. Strategy&. October.

Quijano, A. (2000). "Capítulo: Colonialidad del poder, Eurocentrismo y América Latina." In *Colonialidad del saber y Eurocentrismo*, ed. E. Lander, pp. 201–46. Buenos Aires: Edición UNESCO-CLACSO.

Quintanilla, M. (2007). "La investigación en la sociedad del conocimiento." *Revista CTS*, vol. 3, no. 8, abril de 2007, pp. 183–94.

R&D World. (2020). *2020 Global R&D Funding Forecast*. Special mid-year update. August. www.rdworldonline.com.

Rathenau Instituut. (2020). *European science and innovation in a new geopolitical arena*. The Hague, authors: L. Hessels, S. Y. Tjong Tjin Tai, J. Jansen, and J. Deuten.

Ricardo, D. (1817). *On the Principles of Political Economy and Taxation*. London: John Murray.

Risse, T. (2008). "Transnational Actors and World Politics." In *Handbook of International Relations*, ed. W. Carlsnaes, T. Risse, and B. Simmons. Los Angeles: SAGE Publications.

Robertson, R. (1995). "Glocalization: Time-Space and Homogeneity-Heterogeneity." In *Global Modernities*, ed. M. Featherstone, S. Lash, and R. Robertson. London: SAGE Publications.

———, ed. (2015). *European Glocalization in Global Context*. London: Palgrave MacMillan.

Rodrik, D. (2011). *The Globalization Paradox: Democracy and the Future of the World Economy*. New York: W.W. Norton.

———. (2020). "Why Does Globalization Fuel Populism? Economics, Culture, and the Rise of Right-Wing Populism." *Annual Review of Economics*, September.

Rosenau, J. (1990). *Turbulence in World Politics. A Theory of Change and Continuity*. Princeton, NJ: Princeton University Press.——. (1995). "Governance in the Twenty-first Century." *Global Governance*, vol. 1, no. 1 (Winter 1995), pp. 13–43.

———. (2003). *Distant Proximities: Dynamics Beyond Globalization*. Princeton, NJ: Princeton University Press.

Rozas, P., and R. Sánchez. (2005). *Desarrollo de Infraestructuras y crecimiento económico: revisión conceptual*. CEPAL - SERIE Recursos naturales e infraestructura no. 75, October. Santiago de Chile: Naciones Unidas - CEPAL.

Sassen, S. (2001). *The Global City*. New York: Princeton University Press.

———. (2007). *Sociology of Globalization*. New York: W.W. Norton.

———. (2017). "Embedded Borderings: Making New Geographies of Centrality." *Territory, Politics, Governance*, vol. 6, no. 1, pp. 1–11.

SciTech DiploHub (2019). *The Barcelona Manifesto for a City-Led Science and Technology Diplomacy*. http://www.scitechdiplohub.org/manifesto/.

Schumpeter, J. (1961). *History of Economic Analysis*. New York: Oxford University Press.

Schwab, K. (2016). *The Fourth Industrial Revolution*. New York: Crown Business.

———. (2018). *Shaping the Fourth Industrial Revolution*. Geneva: World Economic Forum.

Scott, P. (2002). "Changing Players in a Knowledge Society." In IAU Conference: *Globalisation: What Issues are at Stake for Universities?* - Université Laval in Quebec City, September 20.

———. (2020). "Universities in a 'Mode 2' Society." In *Missions Of Universities. Higher Education Dynamics*, ed. L. Engwall vol. 55, pp. 95–113. Cham: Springer.

Sebastián, J. (2000). "Las redes de cooperación como modelo organizativo y funcional para la I+D." *Redes*, vol. 7, no. 015, agosto 2000, pp. 97–111.

———. (2004). *Cooperación e internacionalización de las universidades*. Buenos Aires: Biblos.

———. (2010). "Las redes como instrumentos funcionales en el ecosistema de la Educación Superior." *Transatlántica de educación*, no. 8, pp. 87–92.

Shah, D. (2018). "By the Numbers: MOOCS in 2017." *Class Central*. January.

Shanghai Ranking Consultancy. (2020). *The 2020 Academic Ranking of World Universities*. http://www.shanghairanking.com/.

SIPRI. (2018). *World Military Expenditure Grows to $1.8 Trillion in 2018*. April 29. https://www.sipri.org/media/press-release/2019/world-military-expenditure-grows-18-trillion-2018.

Skolnikoff, E. (1993). *The Elusive Transformation: Science, Technology, and the Evolution of International Politics*. Princeton, NJ: Princeton University Press.

———. (2002). "Will Science and Technology Undermine the International System?" *International Relations Asia Pacific*, vol. 2, pp. 29–45.

Slater, D. (2008). "Re-pensando la geopolítica del conocimiento: reto a las violaciones imperiales." *Tabula Rasa*, no. 8, enero-junio, 2008, pp. 335–58.

Slaughter, S., and L. Leslie. (1999). *Academic Capitalism. Politics, Policies, and the Entrepreneurial University*. Baltimore: Johns Hopkins University Press.

Smouts, M. C. (2001). "Las mutaciones de una disciplina." *Política y Cultura*, no. 15, primavera 2001.

Sporn, B. (2006). "Chapter 9: Governance and Administration: Organizational and Structural Trends." In *International Handbook of Higher Education*, ed. J. Forest and P. Altbach. Dordrecht: Springer.

State Council of China. (2015). *Made in China 2025*. http://english.www.gov.cn/2016special/madeinchina2025/.

Stiglitz, J. (2002). *The Globalizations and Its Discontents*. New York: W.W. Norton.

Strange, S. (1986). *Casino Capitalism*. Oxford: B. Blackwell.

———. (1994). *States and Markets*. London: Continuum.

———. (1996). *The Retreat of the State: The Diffusion of Power in the World Economy*. Cambridge: Cambridge University Press.

Tejada, G. (2012). "Mobility, Knowledge, and Cooperation: Scientific Diasporas as Agents of Development." *Migration and Development*, vol. 10, no. 18, 59–92.

Tello Beneitez, M. (2013). *Guía de Think Tanks en España*. Segunda Edición. Madrid: Fundación Ciudadanía y Valores.

The Commission on Global Governance. (1995). *Our Global Neighbourhood*. Oxford: Oxford University Press.

The Economist. (2014). *Coming to an Office near You*. January 18.

The Royal Society. (2010). *New Frontiers in Science Diplomacy. Navigating the Changing Balance of Power*. January. London: The Royal Society Science Policy Centre.

———. (2011). *Knowledge, Networks, and Nations: Global Scientific Collaboration in the 21st Century*. March. London: The Royal Society Science Policy Centre.

Todt, O. (2021). "The Limits of Policy: Public Acceptance and the Reform of Science and Technology Governance." *Technological Forecasting and Social Change*, vol. 78, pp. 902–9.

Toffler, A. (1991). *Powershift. Knowledge, Wealth, and Violence at the Edge of the 21st Century*. New York: Bantam Books.

Ugalde Zubiri, A. (2006). "La acción exterior de los Actores Gubernamentales No Centrales: un fenómeno creciente y de alcance mundial." *Politika. Revista de Ciencias Sociales*, no. 2, diciembre 2006, pp. 115–28.

UNCTAD. (2020). *A Framework for Science, Technology and Innovation Policy Reviews: Harnessing Innovation for Sustainable Development.* Geneva: United Nations.

UNESCO. Data Centre. The UNESCO Institute for Statistics. http://data.uis.unesco.org/.

———. (2005). *Towards Knowledge Societies.* Paris: UNESCO.

———. (2006). *Sixty Years of Science at UNESCO 1945–2005.* Paris: UNESCO.

———. (2009). *Conferencia Mundial de Educación Superior: Nuevas dinámicas de la Educación Superior y de la Investigación para el Cambio Social y el Desarrollo.* (Comunicado - 8 de julio de 2009). Paris: UNESCO.

———. (2010a). *Science Report 2010: The Current Status of Science around the World.* Paris: UNESCO.

———. (2010b). *World Social Science Report 2010: Knowledge Divides.* Paris: UNESCO.

———. (2015). *UNESCO Science Report: Towards 2030.* Paris: UNESCO.

———. (2017). *Women in Science.* UIS Fact Sheet no. 43, March.

Union of International Associations (UIA). (2014). *Yearbook of International Organizations 2016-2017. Volume 5 - Statistics, Visualizations, and Patterns.* Brussels: UIA and Brill.—

———. (2020). *Yearbook of International Organizations 2020–2021.* Brussels: UIA and Brill.

Van Dijk, J., and K. Hacker. (2003). "The Digital Divide as a Complex and Dynamic Phenomenon." *The Information Society*, vol. 19, pp. 315–326.

Van Dijk, J.. (2020). *The Deepening Divide: Inequality in the Information Society.* 4th ed. Los Angeles: SAGE.

van Kersbergen, K., and F. van Waarden. (2004). "'Governance' as a Bridge between Disciplines: Cross-disciplinary Inspiration Regarding Shifts in Governance and Problems of Governability, Accountability and Legitimacy." *European Journal of Political Research*, vol. 43, no. 2, March, pp. 143–71.

Van Wyk, R. (2004). *Technology: A Unifying Code. A Simple and Coherent View of Technology.* Edina: Stage Media Group.

Vincent-Lancrin, S. (2006). "What Is Changing in Academic Research? Trends and Futures Scenarios." *European Journal of Education*, vol. 41, no. 2, June, pp. 1–27.

Wallerstein, I. (1991). *Geopolitics and Geoculture: Essays on the Changing World-System.* Cambridge: Cambridge University Press.

———, coord. (1996). *Open the Social Sciences: Report of the Gulbenkian Commission on the Restructuring of the Social Sciences.* California: Stanford University Press.

Weiss, C. (2005). "Science, Technology, and International Relations." *Technology in Society*, vol. 27, no. 3, pp. 295–313.

———. (2015). "How Do Science and Technology Affect International Affairs?" *Minerva*, vol. 53, no. 4, pp. 411–430.

WIPO. (2020). *2017 - PCT Yearly Review. The International Patent System.* Genève: WIPO Economics & Statistics Series.

World Bank. (n.d.). *World Development Bank.*Online database. https://data.worldbank.org/ (last accessed November 20, 2020).

World Bank Institute. (2010). *Measuring Knowledge in the World's Economies.* Washington, DC: Knowledge for Development Program.

World Economic Forum. (2011). *Global Talent Risk—Seven Responses.* Genève: World Economic Forum and the Boston Consulting Group.

———. (2018). *The Next Economic Growth Engine: Scaling Fourth Industrial Revolution Technologies in Production. White Paper.* In collaboration with McKinsey & Company.

World Science Forum. (1999). *Declaración Sobre la Ciencia y el uso del Saber Científico.* Adoptada por la Conferencia mundial sobre la Ciencia el 1 de julio 1999. Budapest: UNESCO.

———. (2011). *Declaration of the Budapest World Science Forum 2011 on a New Era of Global Science.* Budapest: WSF.

———. (2017). *Declaration of the 8th World Science Forum on Science for Peace.* Text adopted on November 10. Dead Sea, Jordan.

———. (2019). *Declaration of the 9th World Science Forum on Science: Science Ethics and Responsibility.* Text adopted on November 23. Budapest, Hungary.

Yakushiji, T. (2009). "The Potential of Science and Technology Diplomacy." *Asia-Pacific Review*, vol. 16, no. 1, May, pp. 1–7.Youngs, G. (2004). "Feminist International Relations: A Contradiction in Terms? Or: Why Women and Gender Are Essential to Understanding the World 'We' Live In." *International Affairs*, vol. 80, no. 1, January, pp. 75–87.

Zurbriggen, C., and M. González Lago. (2010). *Análisis de las iniciativas MERCOSUR para la promoción de la Ciencia, la Tecnología y la Innovación.* (Número 007). Montevideo: Centro de Formación para la Integración Regional (CEFIR).

INDEX

academic capitalism 47
academic institutions 138
   international students 138
   world-class universities 137
academic networks 119
actors xiii, 41–70, 199–200
   Epistemic Communities 65–68
   individuals 25
   interactions 72, 123
   Intergovernmental Organizations 53–56
   international system 23–25
   multipolar world 25
   Nation-State 24, 48–53
   new 24–25, 41–42, 45
   Non-Governmental Organizations 57–60
   non-governmental transnational actors 24
   old 41–42
   production process 105
   relations 72
   scientific diasporas 68–70
   think tanks 63–65
   Transnational Companies (TC) 60–62
   universities 42–48
African Union (AU) 84
Age of Discontinuity 14
Alegre, Moreno 89
Altbach, Philip 43
American Association for the Advancement of Science (AAAS) 60
application process of scientific knowledge 138–44, 203–04
   private sector 143–44
   publications and patents 113–14
   research, type of 139–40

   scientific fields 140–41
   socioeconomic impact 142–43
applied research 139
artificial intelligence (AI) 17, 144, 174, 177
Association of Southeast Asian Nations (ASEAN) 56, 84
asymmetric relations 97–100
   cognitive divide and dependency 99–100
   inequalities 97–98
   knowledge 98–99
   power 98–99

Bacon, Francis 4
basic research 139
Bayh-Dole Patent 80
Beneitez, Tello 63
bifurcated global structure 25
Big Science 10, 49–50, 103
bionic industry 176
biotechnology 176
blogs 124
Bokova, Irina 83
Bologna Process 85, 151
BRAC 57
brain drain 78
brain gain 78
brain network 69
Breton, Gilles 46
BRICS 172
Bull, Hedley 25
Bunge, Mario 6

Calame, Pierre 24
capitalist economic system 34–35
Caribbean Community (CARICOM) 56
Casino Capitalism 38

INDEX

Castells, Manuel 148
Cervera, Calduch 72
change macrotrends 217–18
China xiv, 173–74
cognitive divide 160
Cold War, end of 30–31
collaboratories 119
commercialization of STI 166–68
   scientific knowledge production, methods of 167–68
   societal relevance, decline of 168
   value 166–67
communications xiii
competitive interactions 28, 90–97
   between companies 91–93
   between states 90–91
   for new emerging technologies 95–97
   university rankings 93–94
complex interdependence 26, 27
conflictive interactions 73–82
   global talent 73–79
   intellectual property rights and patents 79–82
conflictive links 28
continuity macrotrends 217
cooperation 28, 82–90
   international 83–85
   inter-university 89–90
   scientific 83
   States, companies, and universities, links between 87–88
   STI diplomacy 85–87
cooperation networks 118–19
corporate universities 62
COVID-19 pandemic 121
Cox, Robert 29, 188
cyberattacks 181
cybercrime 181
cyberespionage 182
cybersecurity 181, 182
cyberspace 180
cyberspace governance 182–83
cyberterrorism 182

de Semir, Vladimir 125
Dedijer, Stevan 75
democratizing element in society, scientific knowledge as 192–94
Deudney, Daniel 12
digital devices 120
digital divide 159
Digital Economy 95
digital media 125–27
   immaterial diffusion 126
   network diffusion 126
Discourse on the Method 4
distribution process, ISR 130–38, 203
   academic institutions 138
   geoeconomic 130–31
   geopolitical 130–31
   patents 135–36
   postgraduate students 138
   R&D investment 131–33
   researchers 133–34
   scientific publications, geographic distribution of 134–35
Doctors Without Borders 57
drones 17
Drucker, Peter 14

Echeverría, Javier 167
Economic Community of West African States (ECOWAS) 56
economic globalization 37
economic system 212–13
economic wealth, resources for 185–87
economy 61
e-estonia project 214
E-learning 119–21
electronic devices 120
emerging realities 157–83, 204–05
   gender divide 169–71
   geopolitical changes 172–75
   knowledge gap 158–62
   multilevel and global governance 162–66
   new emerging phenomena 158
   science and emerging technology 175–80
emerging technologies 95–97
Epistemic Communities 65–68
   expert knowledge, revaluation of 67–68
   in ISR 66
   role 67
   traits 66
Europe 9

INDEX

European Commission's Framework Program (FP) 151
European Research Area (ERA) 151
European Strategy Forum on Research Infrastructures" (ESFRI) 113
European Union (EU) 55, 84, 150
experimental development 140
explicit/codified knowledge 115

Federal Technology Transfer Act (1986) 80
Fernández, Albáizar 89
formal education 115–21
Foucault, Michel 43
Fourth Industrial Revolution 95
future governance, STI 165–66

gender divide 169–71
General Agreement on Trade in Services (GATS) 55
geoeconomic 130
geoeconomic distribution process 130–31
geopolitical changes 172–75
   and geoeconomic shift 174–75
   China's rise 173–74
   regional changes 172–73
geopolitical distribution process 130–31
geopolitics 130
Glasnost (political reform) 30
global agenda 27
global brain chain 78
global city 148
Global Context 66–67
global governance 164–65
global interdependence 36
global structure, ISR 205–10
   hierarchization 208–09
   polarization 205–07
   scientific knowledge, geopolitical and economic distribution of 207–08
   sectorization 210
   segmentation 209
global talent 73–79
   historic evolution 75
   impact 78–79
   public policies 76–78
   scarcity 75–76
global village 33
globalization 35–38

glocalization 148
Gorbachev, Mikhail 30, 31
governance process 144–53, 204
   international arena 151–53
   local governments 148–49
   public policies 145–48
   regional initiatives 150–51
Gramsci, Antonio 29
Guardilla, García 99

Haas, Peter 66
Haass, Richard xv, 36, 38
Hagel, Chuck 197
Halliday, Fred 30
hard power 86, 188
hegemony 29
hierarchization 208–09
higher education 110–11
highly skilled personnel 128–30, *See* global talent
   global impact 129–30
   movement dynamics 128–29
Horkheimer, Max 188
hybrid conflicts 183

Immigration Act of 1990 76
immigration law 76
Industrial Revolution 9
inequalities 97–98
information and communication networks 118
information and communication technologies
   digital media 125–27
   mass media 125
Information Society 16
infrastructures 112–13
Innerarity, Daniel 161
innovation hubs 53
innovation networks 119
Innovation Systems 104
innovations 6, 15
institutionalism 62
intellectual property rights and patents 79–82
   evolution 80–81
   interests, clash of 81–82
   origin 79–80

INDEX

interactions 26–29, 71, 201, 212–16
  actors 72
  breeding ground 72
  competitive 90–97
  conflictive 73–82
  global agenda 27
  interdependence 26
  linkages, types of 28–29
  mechanism 71–73
  production process 105
  states 26
  with economic system 212–13
  with political system 213–14
  with social system 215–16
  with strategic-military system 214–15
interdependence 26
Intergovernmental Organizations (IGOs) 53–56
  regional processes 55–56
  Specialized Organizations 54–55
intermediation process, ISR 114–30, 202–03
  cooperation networks 118–19
  E-learning 119–21
  formal education 115–21
  higher education 110–11
  highly skilled personnel, mobility and circulation of 128–30
  knowledge and technology transfer 121–24
  lifelong learning 117–18
  scientific research 116–17
international arena 151–53
international cooperation 83–85
International Council for Science 59
International Organization for Science and Technology Education (IOSTE) 60
international processes 30–38
  capitalist economic system, changes in 34–35
  Cold War, end of 30–31
  globalization 35–38
  Nation-States, challenges to 31
  Scientific and Technological Revolution 32–34
international relations, STI and 8–13
  disciplinary approach 10–13
  practical links 8–10

international scientific relations (ISR) xviii, 17–20
  actors 199–200
  and international system 211–21
  challenges 19–20
  configuration 199–210
  democratizing element 192–94
  economic wealth in 185–87
  emerging realities 157–83, 204–05
  empirical phenomenon 17–18
  global configuration 226–30
  global structure 205–10
  in military field 194–97
  interactions 201, 212–16
  international system and 230–31
  macrotrends 216–18
  power, instrument of 187–89
  processes 101–53, 201–04
  relations 201
  roles of scientific knowledge in 185–97
  social innovation, mechanism of 189–92
  subdisciplinary field 18–19
  systemic parameters 199–205
  university in 42
international students 138
international system 23–40
  actors 23–25
  economic system 212
  global agenda 27
  global configuration of 38–40
  implication for 218–21
  interactions/relations within 26–29
  intersystemic transition 38–39
  ISR and 230–31
  linkages, types of 28–29
  macrotrends to 216–18
  multipolar world 25
  Nation-States 24
  new actors 24–25
  new global configuration 39–40
  political system 213–14
  processes 30–38
  social system 215–16
  strategic-military system 214–15
Internet 127
intersystem transition 38–39

inter-university cooperation 89–90
inventions 15

Jover, Núñez 107

Kaeser, Joe 177
Kant, Immanuel 28
Kennedy, Paul 8
Keohane, Robert 24, 26
King, Roger 45
knowledge and technology acquisitions 123
knowledge and technology transfer 121–24
   actors and intervening interactions 123
   digital media for 125–27
   information and communication technologies, diffusion through 124–27
   mass media, dissemination by 125
   nonlinear processes 123
   process of 122–24
   relevance of 121–22
   sources 123
Knowledge Economy 14–16, 61, 121, 186
knowledge gap 158–62
   cognitive divide 160
   digital divide 159
   divides, impact of 161–62
   intra-societal differences and 161–62
   new centers/peripheries and 161
   scientific divide 160
   social divides 162
   types 158–60
Knowledge Society 16, 116, 160
Krepinevich, Andrew 182

Leslie, Larry 47
lifelong learning 117–18
linear model 49
linkages, types of 28–29
local governments 148–49

Machlup, Fritz 14
macrotrends, ISR 216–18
   change 217–18
   continuity 217
Manhattan Project 49
Mariscal, Nicolás 145
mass media 125
Massive Open Online Courses (MOOCs) 120

microprocessor 33
Migration for Development in Africa (IMO) program 70
military field, scientific knowledge 194–97
modern science 4
Morata, Francesc 145
Moyua, Celestino del Arenal 24
multilevel governance 163–64
multipolar world 25

nanotechnology 176
National Innovation Systems (NIS) 103
Nation-State 24, 41, 48–53
   and scientific knowledge 49
   Big Science 49–50
   in international political system 31
Nation-States 144
   challenges to 31
   post–Cold War era 50
   sub-state entities 52–53
Nazism 9
neo-medievalism 25
Network Societies 36, 126
Network Society 127
new actors 24–25, 41–42, 45
new connectivity 148
Newman, John Henry 43
newspapers 124
non-central government actors 52
Non-Governmental Organizations (NGOs) 57–60
   field of operation 58
   in STI 59–60
   rise of 57–58
   scope of application 58
   STI, promotion and uses of 58–59
non-governmental transnational actors 24
non-State actors 24
Novum Organum 4
Nye, Joseph 24, 26

online communication 120
open information sources 123
Organization for Economic Co-operation and Development (OECD) 55, 84
Oslo Manual 6
Oxfam International 57

patent 62, 113–14, 135–36
Patent Cooperation Treaty (PCT) 81
Peace of Westphalia 24
Pedraza, Dallanegra 38
Perestroika (economic reform) 30
perpetual peace 28
places of production 104
Polanyi, Michael 115
polarization 205–07
political system 213–14
postgraduate students 138
Post-Normal Science model 106
power of knowledge 187–89
private sector 143–44
processes, ISR 101–53, 201–04
   application of scientific knowledge 138–44, 203–04
   distribution 130–38, 203
   governance 144–53, 204
   intermediation 114–30, 202–03
   production 102–14, 202
production process 102–14, 202
   actors 105
   alternative methods 107–08
   application 105
   factors 108–14
   higher education and 110–11
   infrastructures 112–13
   interactions 105
   levels 104
   methods 102–08
   new methods 104–07
   places 104
   publications and patents 113–14
   quality control 106
   R&D 109–10
   researchers in 110
   results 106
   social responsibility and reflexivity 105
   traditional methods 103–04
   transdisciplinarity 106
Project Maven 96
public policies 145–48, 171
publications and patents
   R&D 113–14

Red Caldas 70
regional initiatives 150–51
regional processes 55–56
relations 26–29, 71–100, 201
   actors 72
   asymmetric 97–100
   competitive interactions 90–97
   conflictive interactions 73–82
   cooperation 82–90
   global agenda 27
   interdependence 26
   linkages, types of 28–29
   mechanism 71–73
   network of 71
   states 26
Research and Development (R&D) 10
   distribution process, ISR 131–33
   in civil 142
   in defense 142
   production process 109–10
   socio-economic objectives 142
research networks 119
researchers 110, 133–34
Reunión de Ministros y Altas Autoridades de Ciencia, Tecnología e Innovación (RMACTIM) 151
Reunión de Ministros y Altas Autoridades de la Ciencia, Tecnología e Innovación (RMACTIM) 56
Reunión Especializada de Ciencia y Tecnología (RECYT) 56, 151
Risse, Thomas 25
robotics 176
Roosevelt, Franklin D. 49
Rosenau, James 25, 145
Royal Iron Works in Weald (Sussex) 9
Royal Society 59

Sánchez, Castro 107
Santos, Boaventura de Sousa 45, 94
Sassen, Saski. 161
Schumpeter, Joseph 6
Schwab, Klaus 34
science 5
science and emerging technology 175–80
   future implications 177–80
   new areas 176–77
   virtual world 180–83
Science, Technology, and Innovation (STI) xiii, xvii–xxii, 42

and international relations xvii, 3, 8–13
as empirical phenomenon 5–7
as subject of study 7–8
as tool 59
China in 173–74
commercialization 166–68
diplomacy 85–87
future governance 165–66
impulse and promotion of 58
in NGOs 58–59
Knowledge Economy 14–16
Knowledge Society 16
multilevel and global governance 162–66
on society 7–8
revaluation of xix
rise of 13–16
role of 224–26
science, dawn of 4–5
scientific fields, States investments in 140
scientific knowledge, new role of 13–14
women in 169–71
Scientific and Technological Revolution xv, 32–34
scientific cooperation 83
scientific diasporas 68–70
  from brain drain to brain gain 69
  initiatives 69–70
scientific divide 160
scientific fields 140–41
scientific knowledge 4, 5, 7, 8, 50
  and international relations 9
  and Nation-State 49
  application of 138–44
  as democratizing element 192–94
  commercialization of 166–67
  for economic wealth 185–87
  governance 144–53
  in international system 8–13
  in military field 194–97
  intermediation process for 114–30
  new role of 13–14
  power, instrument of 187–89
  production of 102–14, 167–68
  roles of 185–97
  social innovation, mechanism of 189–92
  tacit vs.explicit/codified knowledge 115
  universities as actor of 43

scientific policies 146
scientific publications 113–14, 134–35
scientific research 62, 116–17
Scott, Peter 44
Second Industrial Revolution 9
sectorization 210
segmentation 209
Silicon Valley 53
Skolnikoff, Eugene 14
Slater, David 98
Slaughter, Sheila 47
smartphones 33
social divides 162
social innovation, mechanism of 189–92
social media 124
social system 215–16
Society 5.0, Japanese plan 216
soft power 86, 188
South African Network of Skills Abroad (SANSA) 70
Southern Common Market (MERCOSUR) 84, 151
South-South Cooperation 85
Soviet Union 10, 30
Specialized Organizations 54–55
Sporn, Barbara 44
state actors 25
state-of-the-art technology xiii
STI Diplo 85
Strange, Susan 25
strategic-military system 214–15
structural power 188
sub-state entities 52–53, 149
Sustainable Development Goals (SDGs) 27
System 0 xv

tacit knowledge 115, 128
talents 61
  global 73–79
  mobility 74
Taylor, Frederick Winslow 14
technology 6, *See also specific technologies*
technology transfer 62
Technology Transfer Offices 116
television 124
Theory of Innovations 6

think tanks 63–65
  autonomy, margins of. 65
  challenges 65
  distribution/geographical expansion 64–65
  predecessors 63–64
Third Mission, university 116
Toffler, Alvin 188
Trademark Laws Amendment (1980) law 80
Trade-Related Aspects of Intellectual Property Rights (TRIPS) 80
training 118
Transfer of knowledge through expatriate nationals (UN) program 70
Transnational Companies (TC) 60–62
  action strategies 61–62
  economic interest 61
  STI in 60
Triadic patent 99
Triple Helix model 87, 105

Union of South American Nations (UNASUR) 56
United Kingdom 9
United Nations Educational, Scientific, and Cultural Organization (UNESCO) 54, 84
United Nations General Assembly 27
United States xiv, 9, 30, 197
Universal Declaration of Human Rights 81
universities 42–48, 123, 137
  as actor of scientific knowledge 43
  as universal vocation 43
  autonomy 46
  capacity 44
  challenges 45–47
  crisis of hegemony 45
  flexibility 44
  historical particularities 43–45
  institutional crisis 46
  new actors 45
  new roles 45–47
  reconfiguration/adaptation 47–48

virtual conflicts 181–82
virtual world 180–83
  cyberspace governance 182–83
  development and cooperation 180–81
  virtual conflicts 181–82
Viterale, Francisco del Canto xv

Wallerstein, Immanuel 45
websites 124
Weiss, Charles 12
Westfailure system 32
Westphalian order 32
women in STI 169–71
  public policies 171
  statistical analysis of 170–71
World Intellectual Property Organization (WIPO) 55, 80
World Trade Organization (WTO) 55, 80
world-class universities 94, 137

Lightning Source UK Ltd.
Milton Keynes UK
UKHW010415130821
388772UK00001B/21